BEYOND PEDAGOGIES OF EXCLUSION IN DIVERSE CHILDHOOD CONTEXTS

CRITICAL CULTURAL STUDIES OF CHILDHOOD

Series Editors:
Marianne N. Bloch, Gaile Sloan Cannella, and Beth Blue Swadener

This series will focus on reframings of theory, research, policy, and pedagogies in childhood. A critical cultural study of childhood is one that offers a "prism" of possibilities for writing about power and its relationship to the cultural constructions of childhood, family, and education in broad societal, local, and global contexts. Books in the series will open up new spaces for dialogue and reconceptualization based on critical, theoretical, and methodological framings, including critical pedagogy, advocacy and social justice perspectives, cultural, historical, and comparative studies of childhood, and post-structural, postcolonial, and/or feminist studies of childhood, family, and education. The intent of the series is to examine the relations between power, language, and what is taken as normal/abnormal, good and natural, to understand the construction of the "other," difference and inclusions/exclusions that are embedded in current notions of childhood, family, educational reforms, policies, and the practices of schooling. *Critical Cultural Studies of Childhood* will open up dialogue about new possibilities for action and research.

Single-authored as well as edited volumes focusing on critical studies of childhood from a variety of disciplinary and theoretical perspectives are included in the series. A particular focus is in a reimagining as well as critical reflection on policy and practice of education in early childhood and at the primary and elementary levels. It is the intent of this series to open up new spaces for reconceptualizing theories and traditions of research, policies, cultural reasonings, and practices at all of these levels, in the United States, as well as comparatively across all these levels.

Published by Palgrave:

The Child in the World/The World in the Child: Education and the Configuration of a Universal, Modern, and Globalized Childhood
Edited by Marianne N. Bloch, Devorah Kennedy, Theodora Lightfoot, and Dar Weyenberg; Foreword by Thomas S. Popkewitz

Beyond Pedagogies of Exclusion in Diverse Childhood Contexts: Transnational Challenges
Edited by Soula Mitakidou, Evangelia Tressou, Beth Blue Swadener, and Carl A. Grant

"Race" and Early Childhood Education: An International Approach to Identity, Politics, and Pedagogy
Edited by Glenda Mac Naughton and Karina Davis

BEYOND PEDAGOGIES OF EXCLUSION IN DIVERSE CHILDHOOD CONTEXTS

Transnational Challenges

Edited by

Soula Mitakidou, Evangelia Tressou,
Beth Blue Swadener, and Carl A. Grant

First published in 2009 by
PALGRAVE MACMILLAN®
in the United States—a division of St. Martin's Press LLC,
175 Fifth Avenue, New York, NY 10010.

Where this book is distributed in the UK, Europe and the rest of the world, this is by Palgrave Macmillan, a division of Macmillan Publishers Limited, registered in England, company number 785998, of Houndmills, Basingstoke, Hampshire RG21 6XS.

Palgrave Macmillan is the global academic imprint of the above companies and has companies and representatives throughout the world.

Palgrave® and Macmillan® are registered trademarks in the United States, the United Kingdom, Europe and other countries.

ISBN: 978–0–230–61284–6

Library of Congress Cataloging-in-Publication Data

    Beyond pedagogies of exclusion in diverse childhood contexts : transnational challenges / edited by Soula Mitakidou ... [et al.].
        p. cm.—(Critical cultural studies of childhood)
    Includes bibliographical references and index.
    ISBN 0–230–61284–9
     1. Educational equalization—Cross-cultural studies. 2. Discrimination in education—Cross-cultural studies. 3. Children of minorities—Education—Cross-cultural studies. 4. Marginality, Social—Cross-cultural studies. I. Mitakidou, Christodoula.

LC213.B49 2009
379.2′6—dc22   2008054871

A catalogue record of the book is available from the British Library.

Design by Newgen Imaging Systems (P) Ltd., Chennai, India.

First edition: July 2009

10 9 8 7 6 5 4 3 2 1

Printed in the United States of America.

We dedicate this book to the many educators throughout the world who create inclusive spaces for learning and implement pedagogies of hope and possibility for all children. Their stories, some of which are reflected in this volume, serve to stimulate the important conversations that inspire us to engage in these transformative pedagogies, so necessary to acquiring a flourishing life in an increasingly challenging and interdependent world.

Contents

FOREWORD

Joseph Tobin

As the editors of this collection and many of the chapter authors point out, building pedagogies of inclusion happens not only in schools but also in families and communities. Efforts to overturn implicit and explicit practices of exclusion are most successful when they combine the efforts of teachers, students, parents, and community members. This book is exemplary and unusual among books published on education in that it attends to each of these stakeholders and to their interconnections. There are essays in this collection on curricular and pedagogical practices carried out in schools; essays that emphasize the importance of listening to the voices of parents of students who live in historically excluded and disenfranchised communities; and essays that emphasize the importance of community involvement and even of community control over what happens in classrooms.

These essays, collectively, raise complex questions about the interplay of the local and the global in the production and experience of pedagogies of exclusion. Globalization, though a relatively new term, is not a new phenomenon. As many of the authors in this collection point out, contemporary problems of exclusion have deep historical roots in global phenomena such as slavery and colonialism. Contemporary forms of the global circulation of capital, policies, and people exacerbate old problems and create new ones. But they also can create new opportunities and introduce new ideas and possibilities. When reading the essays in this collection, I was struck by the similarities across the wide-sprung and diverse settings discussed in this book, in both the nature of the problem of educational exclusion and of the solutions to be found to this problem. The range and diversity of settings discussed in this book are impressive, including South Africa, Brazil, Australia, Greece, Argentina, Mexico, the United States, and the Navajo Nation. Some of the papers are about educational challenges facing indigenous communities, some about stigmatized racial groups, and some are about immigrants and refugees. And yet a surprising similarity of problems and solutions emerges.

But to point out this similarity is not to imply that these problems and solutions are not local. Rather, I think the lesson here is that these globally circulating problems have specific local meanings and manifestations. An implication is that while we can borrow under-standings and strategies from one setting and attempt to apply them in another, when we do so we need to remember that the most suc-cessful solutions are successful because they are grounded in local communities and local and indigenous epistemologies, values, and practices. The universal truth here is that the most successful solu-tions are local. A virtue of the essays in this collection is that each is based on close connection to and intimate knowledge of the way inclusion and exclusion play out in a particular local community, a specificity of analysis that allows readers to be introduced to a variety of unique local situations while also having the chance to compare and contrast across the sites.

In the introduction, the editors make a compelling case for the timeliness of this collection and the urgency of the problem by outlin-ing the impacts on classrooms, teachers, students, and their families of neoliberalism and other contemporary globally circulating regimes of power. In order to understand and counteract contemporary forms of exclusion, it makes sense, as most of the authors in this collection have done, to focus on contemporary sources of the problem. These essays demonstrate that in many educational settings in many parts of the world, students are being excluded in many ways. This raises a question that is provocative to contemplate but difficult to answer: are problems of exclusion in education worse today than they were in earlier eras? Because most of these essays are based on analyses of contemporary phenomena and are not historical studies, this is not a question this book addresses directly. In my own recent work doing a sequel to our *Preschool in Three Cultures* study published in 1989, we have struggled with the challenge of thinking about continuity and change in educational beliefs and practices over time. In the new book, *Preschool in Three Cultures Revisited* (2009), I argue, along with my colleagues Yeh Hsueh and Mayumi Karasawa, that it is difficult to steer clear of the dual dangers of narratives of nostalgia and progress. Raymond Williams, in *The Country and the City*, points out that in each generation the (rural) past is idealized as being a more innocent, not-yet-ruined version of the (urban) present, leading to nostalgia for a prelapsarian time and place that existed before things fell apart. In some cases, as, for instance, is especially true for the experiences of Native Hawaiians, Maori, Navaho, and other indigenous peoples prior to colonization, such nostalgia is justified. But in postindustrial societies there is little evidence to support narratives that posit the superiority of social conditions of the recent past as superior to the present (e.g., of the industrial to the postindustrial period). Pedagogies of exclusion exacted a high price on students and society then as now. The opposite problem to nostalgia for the past is idealization of the present and the belief that our societies and their practices are continuously becoming more rational, efficient, and progressive. The authors of this collection challenge the notion that globalization and neoliberal rationality reflect a march toward progress. But it is harder to keep a historically relativistic perspective on our own sense that we are making progress in the development of, well, progressivism, by which I mean the notion that we are getting better, even if still far from perfect at solving social problems such as educational exclusion and bias. As a cultural anthropologist and a relativist, I am suspicious of arguments that posit the superiority not only of one culture over another but also of one time period over another. Some ideas and practices are better and others worse, both practically and ethically, but it requires perspective to know which is which. As Foucault teaches us, many of the most unfortunate and even draconian social practices of the contemporary era were developed not by tyrants, but by social reformers, the progressive thinkers of their day.

The paradox we face in the helping and teaching professions is that we must act and do the best we can to battle practices of exclusion and other problems even though history may judge us as having pursued strategies that, in retrospect, will turn out to have had counterproductive outcomes and to reflect the limitations, the discursive and conceptual prisons, of our era. Faced with this paradox, all we can do is to be as thoughtful, sincere, and self-doubting as we can be and to make a sort of leap of faith into the maelstrom of problems and contradictions of educational problems and solutions, including the problems of educational exclusion and solutions that promise to promote inclusion. The authors of this collection have offered us as thoughtful, sincere, and nuanced set of discussions of the pitfalls and promises of intervening in the circuits of education exclusion as we could hope to find.

My final thought about this excellent collection has to do with connections between the problems and solutions discussed in these essays and the work of the Children Crossing Borders (CCB) project I direct. CCB is an international, interdisciplinary study of im/migrant parent and practitioner perspectives on early childhood education. Our team of twenty researchers from England, France, Germany, Italy, and the United States has conducted focus-group interviews in these five countries with parents and teachers in preschools having a significant percentage of children of recent im/migrants. Our emphasis was on listening to the voices of immigrant parents, voices that have for the most part been ignored by early childhood education teachers and policy makers. We have found many areas of substantial disagreement between immigrant parents and practitioners in particular preschool settings, as for example, in preschools where newly arrived immigrant parents want more emphasis on academics and less on play than the teachers are willing to provide and of migrant parents who want more attention to sexual modesty and more differentiation of boys and girls roles than their children's preschools consider reasonable or just. Most practitioners say they believe strongly in both progressive (e.g., constructivist) practice and in being culturally responsive, but when the two are in conflict, in most cases constructivism trumps cultural responsiveness.

On the other hand, we also have found many examples of agreement between im/migrant parents and teachers and many expressions of goodwill and a determination from both sides to connect. The second stage of our project is focused on piloting strategies for facilitating less hierarchical dialogue between practitioners and im/migrant parents, as we are arguing that antibias curricula, while necessary, are not sufficient, and that the field of diversity in early childhood education has to add to our emphasis on antibias a new emphasis on what we are calling cultural negotiation. I am proposing that education practitioners and programs must be willing to sit down at the table with parents from immigrant and other historically disenfranchised communities to negotiate, in good faith and without preconditions, about how and what is taught in their classrooms. Early childhood educators, many of whom are already skilled at working with young children and respecting their prior knowledge, must be helped to develop skills of working in a nonhierarchical way with the children's parents. As a field, we must move beyond conventional notions of parent participation as parents conducting fund raisers for the school and helping out in the classroom and parents as recipients of teachers explanations and instruction. Instead we can reconceptualize parent participation as an ongoing process of knowledge, value, and power sharing between students, their families, and educators. The essays in this collection start us down the path in this direction by presenting trenchant analyses of the barriers to nonhierarchical negotiation across cultural and racial divides and offering promising solutions suggesting that students, teachers, schools, families, and communities can, under the right circumstances, develop effective pedagogies of inclusion.

SERIES EDITORS' PREFACE

As series editors for Critical Cultural Studies of Childhood, published by Palgrave Macmillan, we are thrilled with the publication of *Beyond Pedagogies of Exclusion in Diverse Childhood Contexts: Transnational Challenges* as the newest contribution to the series.

The series was developed to offer a critical reconceptualization of educational practices from early childhood through elementary or primary education; it was also meant to reexamine the discourse of education, family, and childhood breaking down borders of age, place, and level of education. The book series encourages multi-theoretical, interdisciplinary, and transnational and comparative perspectives on childhood, family, and education in and out of traditional "modern" schooling. It is intentionally oriented toward a critique and reconceptualization of current policies, research, theoretical frameworks, and practices.

From these perspectives, *Beyond Pedagogies of Exclusion in Diverse Childhood Contexts: Transnational Challenges* is an exceptional new contribution to our series. The edited volume beautifully illustrates the ways in which an international group of educational researchers can work together to provide greater clarity and answers to questions related to some of the most important educational issues of today (and tomorrow). The scholarly contributions in this edited book are a result of multiple years of meetings and thus, as an edited volume, also provide coherence and diversity in the ways they examine education and social exclusion as these practices occur in policies, pedagogies, curriculum, and teacher education around the world. With leading scholars from Australia, Greece, South Africa, the United States, and several countries in Latin America, different representations of diversity and exclusion are illustrated, but most importantly, more visionary and inclusive practices are also illustrated, by the majority of the contributors.

The volume's breadth, transnational nature, and focus on various conceptualizations and practices of exclusion, and inclusion, make the volume of interest to policymakers, researchers, teachers, and other educators around the world.

Finally, as the book series on Critical Cultural Studies of Childhood has emphasized throughout, transnational studies of childhood and educational issues help to illuminate new ways of thinking about the language and practices we use that, with even good intentions, eliminate so many from the opportunity for learning, and from participation in social, political, economic, and educational citizenship (as currently constructed) in and across national borders. By emphasizing the influence of neoliberal governance strategies in limiting life chances for many children around the world, as well as the narrow view of competence and educational knowledge embodied in many of our current assimilationist or even "multicultural" policies and pedagogies, the volume helps to open new spaces for discussion, interrogation, and practice.

This volume allows us to refashion notions of diversity, inclusiveness, and learning, through the lens of those working on inclusionary practices in many countries, and provides a vision of what "might work." There are visions and concrete examples of diverse practices to consider and encourage; especially, the book provides a critical lens toward policies and practices that continue to require critique, and, too often, elimination. As we, finally, begin to move further into the twenty-first century with a politics of "hope," a more progressive agenda worldwide is emerging. Again, we are proud to have this book as part of our ongoing discussion—filled intentionally with critique, new ways of seeing things, and new possibilities for thinking, conduct, and action.

Marianne N. Bloch, Gaile S. Cannella, and
Beth Blue Swadener

Acknowledgments

Many have contributed to this volume. We thank all of the presenters and participants at the four conferences that generously gave to the soul of the book. We offer a special thanks to all of the wonderful people at Aristotle University in Thessaloniki, Greece, who made everyone a Greek while they were there. *Efharisto!*—thank you! We offer an elated gratitude to the chapter authors who put up with our deadlines, suggested revisions, and undertook overall editorial responsibilities. We give a special thank you to Ashley Lauren Sullivan for the final formatting and editorial preparation on the manuscript work. Finally, we are very appreciative to Julia Cohen, Palgrave Macmillan Editor Samantha Hasey, editorial assistant, and Erin Ivy, production editor, for their supportive guidance throughout the process of bringing this book to publication.

Introduction

Beth Blue Swadener, Soula Mitakidou,
Evangelia Tressou, and Carl A. Grant

Our children and grandchildren—our future—are living at a time
of rising anti-immigrant and racist social conditions in Europe, the
United States, Australia, and other Western countries. In addition,
tensions exist over migration between poor "sending" countries
and richer "receiving" nations and over several waves of reform and
contestations of education policy globally. This is not to say that all
socially excluded children and families would readily be labeled as
belonging to minority, immigrant, or nondominant cultural backgrounds. Rather, such individuals may not conform to mainstream
trends and pressures and are often, though not always, from low-
income groups. All of this is taking place in a world that is more
connected than ever, while at the same time many parts of the world
are splintered along religious, economic, gender, and racial lines, with
social exclusion growing, in spite of many years of human rights and
multicultural educational research and advocacy.

The United Nations population division predicts that at least
2.2 million migrants will immigrate to wealthier nations every
year till 2050, with an "unprecedented global upheaval" predicted
(*Brussels Journal*, 2007). Large numbers of guest workers, refugees,
and immigrants have become the target of an array of policy initiatives that seek to govern their integration into various societies.
Tensions between Judeo-Christian dominant cultures and Islamic
communities have also added to the dynamics of social exclusion.
The neocolonial immigration of people from Third World nations to
the centers of various former/persisting empires has also led to a rise
in xenophobia and racism, as evidenced from the 2006 riots in the
Paris suburbs and other immigrant rights' protests across Europe. As
unemployment rises and a discourse of post-September 11 antiterror

vigilance and "disaster capitalism" (Klein, 2007) prevails, social, religious, and racial tensions often lurk just below the surface for groups at the margins of established Western societies.

Bringing immigrants, refugees, and long-marginalized groups, such as the Rom and religious minorities, more fully into civil and social society—including providing them, and partnering with them to create, culturally relevant education—is truly a global challenge. Typically, schools are expected to facilitate nation-building and integration through assimilation of such groups. This is complicated further when assimilationist programs are labeled in more inclusive ways.

Approaches to the education of immigrants or migrants contrast greatly from that of the mainstream. In addition to elementary and secondary contexts, children in immigrant families are excluded, or best underserved, in early childhood education and care (Garcia, 2001). When such groups are included, there are often few accommodations made to their specific needs, with an emphasis on rapid language learning and assimilation into the dominant culture. In some cases there are alternative educational programs for particular groups (e.g., Rom schools, Madrassas, heritage language and culture maintenance schools, etc.), but it is still all too rare for these communities to be fully involved, or even consulted, in designing educational programs to serve their children and communities.

The metaphor of Europe as a fortress that shuts illegal immigrants and refugees outside its gates is frequently used. Around 25 million nonnationals (persons who are not citizens of the country in which they reside) were living in the European Union (EU25) Member States in 2004 according to estimates published by Eurostat, the Statistical Office of the European Communities. This represents just fewer than 5.5 percent of the total population of the EU25. Due to differences in concepts, definitions and data sources, and varying rules on the acquisition of citizenship, the international comparability of figures on nonnationals are limited to some extent. However, some observations may be made on the basis of available data.

In 2004, Luxembourg had by far the highest proportion of nonnationals (38.6 percent), followed by Latvia (22.2 percent) and Estonia (20.0 percent) (Eurostat, 2007). In no other Member State was the proportion of nonnationals more than 10 percent. In twelve Member States, nonnationals comprised less than 5 percent of the population. The majority of nonnationals living in Member States were citizens of non-EU countries. However, in Luxembourg (with Portuguese being the largest group), Belgium (Italians forming the largest group), Ireland (Britons forming the largest group), and Cyprus (Greeks comprising the largest group), nonnationals were predominantly citizens of other EU Member States.

The proportion of nonnationals has grown in almost all Member States. The most significant increases between 1990 and 2004 were observed in Luxembourg (from 28.7 percent to 38.6 percent), Greece (from 1.4 percent to 8.1 percent), Spain (from 1.0 percent to 6.6 percent), Cyprus (from 4.2 percent in to 9.4 percent), Ireland (from 2.3 percent to 7.1 percent), and Austria (from 5.7 percent to 9.4 percent). The percentage of nonnationals fell over the same period in Belgium (from 8.9 percent to 8.3 percent), while Latvia (from 27.3 percent to 22.2 percent) recorded a significant decrease of nonnationals between 1998 and 2004 (Eurostat, 2007).

In recent years, it has been found that "migrants" to other nations comprise 2 percent of the world's population, and 20 percent of these migrants are living in the United States. Since 2005, close to 200 million people, or about 3 percent of the world's population, have lived outside their country of birth (*New York Times*, June 24, 2007).

In the United States, immigrants have increased from 10 million in 1970 to over 34 million in 2005. Estimates from the U.S. Census Bureau (2005) predict that their number will reach 42–43 million by 2010. First-generation immigrants comprise 12 percent of the total population and their children make up 22 percent of the school-age population (Urban Institute, 2005), with 53.3 percent of the immigrants being in their childbearing years, and children of immigrants comprising the fastest growing population category in the United States. A number of states have passed English-only legislation or have banned bilingual instruction through ballot initiatives and state and local legislation. The backlash against "illegal" or undocumented individuals and families in the United States is increasing and reflects the increasing xenophobia, economic panic, and continued racism of earlier anti-immigrant movements.

In spring 2006, immigrants' rights movements in the United States came together for several large demonstrations in cities across the country, including San Francisco, Los Angeles, Phoenix, Chicago, New York, and many smaller cities. A problematic wall at the Mexican border is being expanded and federal funding has been provided for heightened security at that border, even as immigration reform legislation languishes in the U.S. Congress. Vigilante groups patrol the Mexican border and counterprotesters monitor such groups. The United States is but one of many nations where social exclusion targets immigrants, migrants, and guest workers.

English as the Language of Globalization

The past two decades have also seen a great expansion of corporate globalization and hypercapitalism, with multinational corporations forming a virtual empire and gaining power that often transcends that of the nation-state. This has been further complicated by the endless "war on terror" and its links to capitalism (Klein, 2007). With globalization, the hegemony of English (Macedo, Dendrinos, and Gounari, 2003) is becoming increasingly apparent—from the rapid growth in English as a Foreign language (EFL) programs, starting in preschool and kindergarten in many countries, to English immersion policies in the United States.

In *The Hegemony of English*, Macedo, Dendrinos, and Gounari, P (2003) point out that the purpose of English language education in the contemporary world order cannot be viewed as simply the development of skills aimed at acquiring the dominant English language. This view sustains an ideology that systematically disconfirms rather than makes meaningful the cultural experiences of the subordinate linguistic groups that are, by and large, the objects of language policies. To quote Macedo and colleagues (2003, 14):

> For the role of English to become understood, it has to be situated within a theory of cultural production and viewed as an integral part of the way in which people produce, transform and reproduce meaning. Thus, the role of English must be seen as a medium that constitutes and affirms the historical and existential moments of lived experience which produces a subordinate or lived culture. It is an eminently political phenomenon and must be analyzed in the context of the theory of power relations and with an understanding of social and cultural reproduction and production.

Sociolinguistic, Cultural, and Political Dynamics of Educational Exclusion

Against a growing global discourse of "Education for All" (EFA), there has been a backlash against various groups and a documented increase in social and educational exclusion for members of nondominant cultural and linguistic groups. While this book focuses, in part, on what it might mean if EFA goals were fully enacted in ways that moved education beyond practices of systematic exclusion, it is also informative to acknowledge the wide gap between policy and practice in focal countries and contexts described herein. From Greece to the Republic of South Africa and to Australia, national policies reflect inclusive education goals, named variously as multicultural, intercultural, cross-cultural, "deracialized," or antiracist. Yet, in every national context discussed by the contributing authors, there are major exceptions to these policies—whether in the persistence of racism in "post-apartheid" South Africa or the increase in anti-immigrant (and anti-indigenous) sentiments in much of Europe, the United States, and—particularly focused on Aboriginal communities—in Australia.

Much of the discourse surrounding immigrants and immigration policy addresses issues such as human rights, economic opportunity, labor shortages and unemployment, the brain drain, multiculturalisms, and integration and flow of refugees and asylum seekers. While education and schooling are central to this discourse, they are often on the periphery in discussion among people outside of the education community. Indeed, education initiatives for nondominant groups (e.g., immigrants, refugees, and indigenous populations) are often marginalized within the larger education community.

A Growing Transnational Dialogue on Pegadogies of Inclusion

As a diverse, interdisciplinary group of scholars, contributors to this volume share an anti-oppressive stance and have utilized an array of theories and research strategies to counter the persistent and growing instances of exclusion in education—based on race, class, gender, ethnicity, language, dis/ability, sexual orientation, and citizenship status. Authors include education researchers, teacher educators, and theorists who have worked for many years on issues of social and educational inclusion and empowerment of groups that have been marginalized.

Some of the contributors to this volume have focused extensively on approaches to creating more democratic, inclusive, and culturally relevant curriculum, focusing on more inclusive strategies of teaching numeracy, literacy, social studies, and science. Others have foregrounded issues of multiculturalism and bi/multilingualism in their work. The contributing authors understand at a personal level the impacts of the demographics discussed earlier and have focused on education policy and related social policies related to ways in which nations govern children and families. All of us share a deep commitment to naming ways in which social exclusion has diminished the educational and life chances of many students in our various sites of work and regions of the world—and to moving the discourse and action beyond pedagogies of exclusion to a more visionary and inclusive praxis.

A major purpose of *Beyond Pedagogies of Exclusion in Diverse Childhood Contexts: Transnational Challenges* is to move the discourse on education from the periphery of discourse on global migration issues by framing a curriculum and an education policy that will enable scholars from several geopolitical contexts to discuss neoliberal and increasingly neoconservative social and educational policy trends in ways that transcend borders. Our hope is to initiate a change-focused dialogue among scholars, practitioners, and consumers of education with an explicit focus on pedagogy. As Giroux (1995, 241) puts it "...it is important to stress that the general indifference of many theorists to the importance of pedagogy as a form of cultural practice does an injustice to the politically charged legacy of cultural studies, one which points to the necessity for combining self-criticism with a commitment to transforming existing social and political problems."

This volume builds upon a series of conferences, workshops, and transnational conversations among an expanding network of scholars, spanning several continents and national contexts. These initial conferences were held in Thessaloniki, Greece, and were facilitated by faculty from Aristotle University's Department of Elementary Education. Faculty from this department have been actively involved in research and community projects focused on countering the social exclusion of children and families (e.g., immigrants, language minority learners, Pontian and Rom community members, etc.) and on antiracist pedagogies across the curriculum. The first conference, held in 2001, focused on literacy and numeracy strategies for language minority students, within a critical multicultural framework. The first two editors of this book were the conference chairs and the second two editors were presenters and co-organizers. Researchers, teacher educators, and classroom teachers from Europe and the United States came together to discuss ideas for addressing educational inequities and social exclusion in schools and communities. Speakers focused on a range of issues from global language loss patterns to issues in mathematics and literacy instruction and raised an array of issues for discussion.

A second conference, again addressing issues of social exclusion and specific curricular strategies, was held in Thessaloniki in 2002, and it included mainly presentations by classroom teachers sharing their concerns and experiences of teaching in multicultural settings. Both of these initial conferences were attended by more than a thousand participants, and their proceedings have been widely referenced.

A third, much smaller workshop/retreat was held in summer 2003 on Aegina, an island close to Athens. This afforded an opportunity for a number of the speakers at the previous conferences to come together in a small group forum for discussion that included ways to reframe "the basics" in education and ways in which more inclusive educational strategies could be used to counter pervasive patterns of social exclusion. Much of this retreat focused on immigrant/migrant/"other" children in schools, and themes from our discussions inform the organization of this book.

In April 2007, a fourth conference was organized by the same department, and parents from minority groups in Greece were invited to share with educators their experiences and struggles with their children's education. Rom, Pontian, and Albanian parents and students joined voices with teachers engaged in antiracist pedagogy to discuss their strategies for taking education beyond pedagogies of exclusion.

Organization of the Book

This book is organized into three sections addressing: (1) theories and policy trends regarding pedagogies and policies of inclusion in a number of national and geopolitical contexts; (2) examples of promising initiatives across the curriculum, which have furthered more culturally and socially inclusive practices; and (3) case studies of innovative teacher education and professional development programs that have promoted inclusionary education.

Several of the chapters in this volume elucidate the nature of children's, parents', and teachers' experiences with language, power, and identity in a variety of geopolitical contexts. In so doing, they uncover the similarities and underlying issues common to children's potentially traumatizing exposure to the large and powerful political and social forces that often serve to exclude and marginalize them. Children and youth who live and grow at the nexus of socio-politico-linguistic conflict may experience the inherently traumatizing effects of exploded identities (Rolstad, Swadener, and Nakagawa, 2008). When children feel ashamed of themselves and their origins, feeling that they have been defined as somehow deviant because of their language and ethnicity, their chances of developing as secure, well-adjusted individuals are greatly diminished (Skutnabb-Kangas, 1981). In her discussion of children's school performance, Skutnabb-Kangas (1981, 235) claims "minority and low SES children tend to have a more modest idea of their abilities than majority and high SES children. This depresses their performance, and the performance in its turn further undermines their self-confidence, and lengthens the odds against them…" As Skutnabb-Kangas and Phillipson (1995) further argue, "In fact, formal education through the medium of majority languages has extremely often forced minority children to assimilate and change identity. We are reminded of the definition of cultural genocide…this transfer can, of course, be either physical or psychological or both" (pp. 72–73).

In the first part, *Pedagogies and Policies of Exclusion and Inclusion*, contributors identify, describe, and critique a number of pedagogies of inclusion that have been applied and have promoted successful practices, situating them in theoretical and sociopolitical contexts. Chapters focus, in part, on policy trends that encourage or impede more inclusive educational practices and highlight particular examples that offer possibilities and hope. Contributors to this section draw from their experiences in inclusive education and critical pedagogy in a number of geopolitical contexts, including Australia, Greece, Brazil, and South Africa.

In the second part, *Curricular Conversations: Successful Programs and Initiatives*, curriculum matters are connected to issues of social exclusion, with emphasis on successful initiatives and programs for students who are often socially and educationally excluded. Given school complexities and challenges, nested in larger national and globalizing ideologies, authors in this section discuss ways in which teaching and learning can go beyond test preparation, so that children acquire not only the necessary skills and content but also learn to take full advantage of knowledge in all aspects of life and in culturally relevant and inclusionary ways. Brief curricular "case studies" offer examples across subject matter areas and geopolitical contexts. Leading curriculum and subject area scholars draw on their research in a number of national contexts, including Japan, Northern Ireland, Argentina (focusing on Bolivian refugees), and with linguistically and culturally diverse educational settings in the United States.

The third part, *Toward Inclusionary Teacher Education and Professional Development*, highlights teacher education and professional development programs that emphasize pedagogies of inclusion. Scholars draw from their work in an array of settings and experiences in Chile, Spain, Australia, the United States, and Kenya. Chapters highlight the use of antiracist, critical, and inclusive practices in pre-service and in-service education, including examples from multicultural/multilingual and indigenous contexts.

The Epilogue looks to the future in our collective work and anticipates further transnational conversations regarding pedagogies of inclusion. This closing chapter brings together the broad themes of the book and advocates a stronger focus on pedagogy in relation to patterns of social exclusion and movement toward more inclusionary policies and practices in education worldwide.

REFERENCES

Brussels Journal (2007). "Millions will Migrate to Europe." (created March 16, 2007, 15:49) Eurostat, 2007. Retrieved on August 20, 2008 from http://www.brusselsjournal.com/node/1982.http://epp.eurostat. ec.europa.eu/portal/page?_pageid=1996,45323734&_dad=portal&_ schema=PORTAL&screen=welcomeref&open=/t_popula/ t_pop&language=en&product=REF_TB_population&root=REF_TB_ population&scrollto=0

Garcia, E. (2001). *Hispanic Education in the United States: Raíces y Alas.* New York: Rowman and Littlefield Publishers.

Giroux H. A. (1995). "Beyond the Ivory Tower: Public Intellectuals and the Crisis of Higher Education." In M. Berube and C. Nelson (eds.), *Higher Education Under Fire: Politics, Economics, and the Crisis of the Humanities.* New York: Routledge. 238–259.

Klein, N. (2007). *The Shock Doctrine: The Rise of Disaster Capitalism.* New York: Metropolitan Books.

Macedo, D., Dendrinos, B., and Gounari, P (2003). *The Hegemony of English.* Boulder, Colorado: Paradigm Publishers.

Makoni, S. and Makoni, B. (2007). "I Am Starving with No Hope to Survive: Southern African Perspectives on Pedagogies of Globalization." *International Multilingual Research Journal*, 1(2), 103–118.

Moll, L. C., Díaz, S., Estrada, E., and Lopes, L. (1992). "Making Contexts: The Social Construction of Lessons in Two Languages." In M. Saravia-Shore and S. F. Arvizu (eds.), *Cross-cultural Literacy: Ethnographies of Communication in Multiethnic Classrooms.* New York: Garland. 339–366.

Rolstad, K., Swadener, B. B., and Nakagawa, N. (2008). "Verde—Sometimes We Call It Green: Construal of Language Difference and Power in a Preschool Dual Immersion Program." *International Journal of Equity and Innovation in Early Childhood*, 6(2), 73–93.

Rosenthal, R. and Jacobson, L. (1968). *Pygmalion in the Classroom.* New York: Holt, Rinchart, and Winston.

Secada, W. and T. Lightfoot. (1993). "Symbols and the Political Context of Bilingual Education in the United States." In M. B. Arias and U. Casanova (eds.), *Bilingual Education: Politics, Practice and Research. 92nd Yearbook of the National Society for the Study of Education Part II.* National Society for the Study of Education: Chicago, Illinois.

Skutnabb-Kangas, T. (1981). *Bilingualism or Not: The Education of Minorities.* (Trans. L. Malmberg and D. Crane). Clevedon, England: Multilingual Matters.

Skutnabb-Kangas, T. and Phillipson, R. (1995). "Linguistic Human Rights, Past and Present." In T. Skutnabb-Kangas (ed.), *Linguistic Human Rights: Overcoming Linguistic Discrimination.* New York: Mouton de Gruyter. 71–110.

PART I

Pedagogies and Policies of Exclusion and Inclusion

Learner Differences: Determining the Terms of Pedagogical Engagement

Mary Kalantzis and Bill Cope

Learning is effective to the extent that it engages with learner identities. These differ deeply and are complex and multilayered. Diversity is at the heart of the educational project, even though, until recently, this has been barely recognized beyond the crudest of age- and grading hierarchies.

This chapter begins by exploring differences in name: what is the range of learner attributes that are relevant to learning? It explores two orders of learner difference: the first comprises what we call "gross demographics," attributes that stare social actors in the face for their conventionality and ordinariness; the second consists of the much more diffuse dynamics of "lifeworld attributes." The chapter then goes on to examine three historically grounded approaches to diversity in modern education: agendas of homogenization, of which exclusion and assimilation are prominent examples; initial phases in the recognition of differences; and then a program that might be more genuinely and effectively inclusive of differences.

The sources of this discussion are practical, political, and theoretical. Practically, the ideas in the chapter are grounded in the work we have done in schools, mostly in Australia but also with colleagues in a number of other countries. Politically, they arise from the Australian context in which Mary Kalantzis has been an activist for immigrant and indigenous rights and where Bill Cope was for a time Director of the Office of Multicultural Affairs in the Department of the Prime Minister and Cabinet, during which time he developed a whole-of-government framework for setting diversity objectives and monitoring them, the *Charter of Public Service in a Culturally Diverse Society* (Department of Immigration and Multicultural Affairs, 1998). Theoretically, the ideas explored in this chapter are based on an evolving conceptual schema, expressed most fully in our *New Learning: Elements of a Science of Education* (Kalantzis and Cope, 2008).

Differences in Name

Difference, Diversity, Divergence

To start with our metaterms—"difference" and "diversity"—what do these terms imply? And to these two metaterms, we want to add a third distinction, "divergence."

Differences are personal or group attributes that stand in contrastive yet functional relation to each other. At the level of gross demographics, we will group these differences into three macro-categories, material, corporeal, and symbolic; at the level of lifeworld attributes, we will consider some of the dimensions of difference that crosscut the demographics. Sometimes differences are neutral with respect to power, hierarchy, symbolism, and access to social resources, though in the radically unequal societies of the past few millennia, this is rarely the case. In these societies, differences mostly embody systematic relations of inequality. All societies have differences. All societies also have endemic ways of dealing with their differences. The way a particular society deals with its differences is a key distinguishing marker of the character and form of that society.

Diversity is the stuff of normative agendas, where differences become the basis for programs of action. Difference, the insistent reality, becomes diversity, the institutional and social response. Many historical and contemporary responses to difference are hardly worthy of the normative descriptor "diversity"—racism, discrimination, and systematic inequity for instance. Nevertheless, these are strategies for managing diversity.

In addition, we want to make another distinction. "Difference" is a found social object. "Diversity" is the mode of recognition of that object. "Divergence" describes a dynamic that is peculiar to some social contexts, such as the societies of "first peoples," and the currently unfolding phase of modernity, in which there is an endogenous, systematic, active, and continuous tendency for individual social agents and groups to differentiate themselves (Kalantzis and Cope, 2006). This is in direct contrast to the earlier modern societies of homogenization or tokenistic recognition of differences.

Naming Differences

Today, a conventional litany of terms is conventionally used to describe and categorize differences—sex, gender, social class, disability, race, ethnicity. Each term, however, is fraught with ambiguities and difficulties. We will call these second order differences, the stuff of gross demographics that stares social actors in the face with a certain kind of obviousness, only to become not-so-obvious on closer examination.

Here are our second order differences:

Second Order Differences: Gross Demographics

Material
  Class
  Locale
  Family
Corporeal
  Age
  Race
  Sex
  Sexuality
  Physical and Mental Abilities
Symbolic
  Culture
  Language
  Gendre (defined as the difficult-to-disentangle symbolic
    representations of gender and sexuality)
  Identity
  Group Affiliation

These are the second order differences that social institutions, such as schools, must negotiate. The dynamics of all these differences have been radically shifting in recent decades and are destined to continue to shift in decades to come. These differences are also becoming more insistent and more demanding of our attention, particularly in long-term and intensive social relationships such as those established in institutionalized education. (See our *New Learning: Elements of a Science of Education.* Cambridge: Cambridge University Press, 2008 for a detailed account of the shifting dynamics of second order differences.)

Difficulties with Gross Demographics

The more insistent the dynamics of difference and the more essential it becomes to the project of managing diversity, the less satisfactory second order categories appear to be. We need to be able to name differences, but as soon as we do, we find that the demographic categories lack the complexity and subtlety we need. We will study two examples first, and then carry out a discussion of the principle.

Example 1: The U.S. Census Category "Race" classifies people into white, black, Hispanic, Asian, and Native American. There are many problems with this demographic categorization, but we will for the moment concentrate on just one, and that is the effect of creating a demographic majority through the construction of the category "White." How is it that Muslim refugees from nineties Bosnia, by fiat of being classed "White," automatically end up in the same category as descendants of the Pilgrim Fathers? If the business of culture or "race" is one of ancestry, then the biggest ethnic group in the United States is people of English ancestry. But at just 14 percent of the population, it would be hard to call Americans of English background an ethnic majority. And if the Amish of Pennsylvania and lapsed Episcopalian gays are both "WASP," this does not have much use as a cultural category, either.

The demographic category of "race" does not describe a physiological or phenotypical reality that has any social significance beyond the construction historically put upon it. In the United States, "race"' in the case of black-white relations is more accurately a marker of the legacy of slavery than of any inherent way humans might tend to react to different skin colors. "Race," in other words, describes a social relation rather than something of biological significance.

What does "race" do, then? The answer, in part, is that it is used to construct a majority, and then differentiate minorities, only circumstantially on the basis of phenotype, but actually on the basis of certain historical experiences. So, it is by no means an accident that many "Whites" (the "majority") can comfortably call themselves unhyphenated Americans whilst "Blacks" are at best hyphenated African-Americans, a "minority" despite the fact that most are descendants of people who first came to the Americas centuries ago and almost as numerous as people of English descent. Similarly, the categories "Hispanic-American" and "Asian-American" are not really about "race," but are markers of certain historical and social relations. The lesson here is that the use of gross demographic categories, like the race categories used in the U.S. census, is deeply loaded and at times profoundly problematic.

Example 2: By the time Bill Cope came to the position of Director of the Office of Multicultural Affairs in the Department of the Prime Minister and Cabinet, the Australian Bureau of Statistics (which administers the census and constructs categories for national data collection) had long abandoned the concept of race. Instead, the Bureau had two major categories of cultural and linguistic ascription, "non-English speaking background" (NESB), and a self and community identified category, "Aboriginal or Torres Strait Islander" (ATSI). In conjunction with the development of the *Charter of Public Service for a Culturally Diverse Society*, the Office of Multicultural Affairs set about developing the concept of "cultural and linguistic diversity" (Department of Immigration and Multicultural Affairs, 1998). Numerous problems were identified with the inconsistent definition and use of the existing categories of NESB and ATSI. They were at times a poor measure of disadvantage, eliding and crudely aggregating differences of class, locale, generation, and the like. They also had the effect of constructing two minorities in juxtaposition to an overly simple, aggregated NESB majority. These categories were much of the time not particularly useful as a tool for policy development or service provision. They also became labels glibly used in bureaucratic shorthand, a kind of perfunctory gesture as if to say that certain fundamental social issues had been addressed because people had been counted, or because a specialist service was available if needed.

As part of the Charter implementation, the acronym "NESB" was abandoned in favor of a general idea of cultural and linguistic diversity that was capable of a more finely grained reading of needs and which could be applied to everyone, not just readily identifiable and nameable minorities. A policy and service framework was established on the basis of different levels of knowledge depending on the nature or intensity of the service encounter. Some encounters do require a great deal of knowledge of the dimensions of cultural and linguistic difference, and for some even any knowledge will suffice. However, longer and more intensive encounters, such as medical and educational services, require a good deal more information than "ATSI" and "NESB" could provide. The Charter proposed that, at a first level of encounter, no questions need be asked. The only issue that arises at this level is the need to know and the level of knowledge that is required or relevant. At the second level, a double-barreled question is asked—country of birth, and, if Australia, whether the person is of Aboriginal or Torres Strait Islander origin. At each of the remaining four levels, one more question is asked: first-language spoken (additional Level 3 question); English language proficiency (additional

Level 4 question); cultural or ethnic background (additional Level 5 question); and religion (additional Level 6 question). In most service encounters, the simple first (Level 2) question is sufficient, country of birth/indigenous status. At the sixth level, all five main questions are asked, and a full cultural and linguistic profile emerges. To this cultural and linguistic profile a dimension of encounter-specific needs can then be added, interests and purposes in which other demographic specifics might interact with aspects of cultural and linguistic diversity, including material and family circumstances, regional location, gender, age, dis/ability, and educational background. Hence, the Charter advocated a matrix approach to reading and addressing cultural and linguistic diversity in a service encounter.

The gross demographics of second order differences do, of course, capture powerful realities. They are rough predictors of educational and social outcomes. For instance, there is some initial value in the tick-box-lists used to identify educationally disadvantaged students.

However, there are major problems with categorization according to gross demographics. First, the business of categorization becomes a series of receding horizons. The more earnest one's application to the task, and the more dedicated and sincere the work of categorization, the more complex the reality that presents itself. The more serious one's focus on second order differences, the finer the distinctions you need to make. Soon, the gross demographics become unwieldy— the subtleties of dis/ability, the catalogue of world cultures, and the proliferation of distinctions of sexuality and gender, for instance.

Second, the internal variation within demographically defined groups is often greater than the average variation between groups. In fact, a rough general metric would be that the internal differences in stance, self-identification and behavior within any demographically defined group is almost always greater than the average difference between groups. This means that the demographic groupings, whilst helpful to our understanding of the historical and experiential basis for certain moral agendas and social claims, are oversimplified and sometimes counterproductively so. Indeed, the categories of gross demographics can easily lead to stereotypical generalizations—about Chinese learning styles, boys' communication styles or the consequences of socioeconomic disadvantage, for instance. For this reason they can prove less than helpful as operational concepts applied to particular individual and group circumstances. As a consequence, diversity programs based on one or several of these categories can at times prove to be ill-judged or irrelevant to individual or subgroup circumstances, or be even counterproductive. In educational contexts, they too easily tend to oversimplify critical success- and failure-determining differences within groups and between individuals. For instance, some students in disadvantaged groups do succeed; background is not all-determining. Indeed, sometimes it is a student's "disadvantaged" background that is the basis for their particular resilience and their peculiar success. Sometimes, also, the gross demographic terms become invidious labels, implying a deficit on the part of the student, when in fact they may be an opportunity upon which to build constructive learning experiences.

Third, the very act of categorization tends to imply that in-group dynamics, cultural attributes, and personal identifications are key. In fact, the categories are all relational, as much a product of the dynamics of the social whole as the product of separable demographics. Class, gender, race, and disability, for instance, are not things in themselves but constructions in and through social relations in which every social actor is implicated. Social groups, in other words, cannot be neatly categorized and described as if that were the beginning and the end of the equity story; rather they are constituted in relationships in which one group or type of person is defined in relation to another. They exist in dynamic and in unstable tension—class to class, gender to gender, culture to culture, ability to ability. They are defined in and through social relationships—of comparative power, privilege, and access to resources. Each group is created through a series of historical and ongoing intergroup relationships. It is these relationships (racism, sexism, comparative socioeconomic privilege, and the like) that often play themselves through in schools and classrooms, and not separate group attributes.

Fourth, differences intersect. They are never things that exist individually. Rather, they are always complex, multilayered realities in which every material, corporeal, and symbolic aspect manifests itself in some way or the other. For any individual the chance of any one particular combination of major, second order demographic dimensions is so low that they invariably find they belong to the tiniest of minorities. Throwing them into one of the larger categories may do disservice to the actual needs and interests of individual people. Again, the second order list is all-too-neat. The groups are not separate; they are overlapping, simultaneous, and multilayered. In fact, virtually every individual represents a peculiar conjunction of dimensions of difference, a unique mix of group or community experiences. The constitution of that individual can only be understood through the subtleties of that individual's narratives of life experience.

Finally, differences never stay still. They are not states that are simply to be found, classified, and dealt with. They are always changing. A shift of emphasis from diversity to divergence means that we are no longer content to leave differences the way they are. We may want to move them along. This can either be from the perspective of an insider—a woman who wants to change the role of women, or an indigenous activist struggling to improve the conditions of life of their people, for instance. Or it can be from an outsider's perspective, such as how we, as educators, assist learners in their self-transformation or growth to achieve dreams and aspirations that may have seemed beyond the scope of possibility within their lifeworlds.

First Order Differences: Lifeworld Attributes

First order or primary differences are the stuff of the lifeworld that give substance to second order or demographic differences. They are the raw materials that give the gross demographics roughly predictable contents that connect abstract demographic categories with concrete patterns of human reality. They also reveal the points at which second order differences overgeneralize or create stereotypes, or when they prove to be unhelpful or plain wrong for particular individuals or subgroups. We call these first order differences "lifeworld attributes."

First Order Differences: Lifeworld Attributes

Life Narratives
   Experiences
   Places of belonging
   Networks
Personae
   Affinities
   Attachments
   Orientations
   Interests
   Stances
   Values
   Worldviews
   Dispositions
   Sensibilities
Styles
   Epistemological
   Learning
   Discursive
   Interpersonal

We need to be able to read these first order differences for the most practical of reasons. Gross demographics are not enough for many critical points of social negotiation, and for particularly intensive, deeply personal, and sustained encounters, such as those created in schools. Not only are the categories too simple for our purposes, but the catalogue of differences we need to have at our disposal in order to be able to read differences in practice becomes more unwieldy as it strives to become more accurate within the second order categorical framework.

The concept of "lifeworld," by contrast, does not set out in the first instance to catalogue substantive differences. (We might, of course, still find this helpful as a second order activity.) The lifeworld consists of the things you end up knowing without having to think how you came to know them. It is the way you end up being without ever having consciously decided to be that way. The lifeworld is not something that is particularly explicit. It is a set of habits, behaviors, values, and interests that go without saying, or are taken for granted, in a particular context. The lifeworld goes without saying because it has come without saying. It is made up of things that seem so obvious to insiders that they don't need to be stated. Knowledge of the lifeworld does not have to be taught in a formal way. You learn how to be in the lifeworld just by living in it, and this learning is mostly so unconscious that it is rarely even experienced as learning. The "lifeworld" is the ground of our existence, the already-learned and continuously-being-learnt experience of everyday life. It is also the locus of our subjectivity and identity, the source of our motivation, and the basis of our agency. It is intuitive, instinctive, and deeply felt.

In a formal educational context, the lifeworld is the everyday lived experience that learners bring to a learning setting. It is the person they have become through the influence of their family, their local community, their friends, their peers, and the particular slices of popular or domestic culture with which they identify. It is a place where the learner's everyday understandings and actions seem to work, such that their active participation is almost instinctive—something that requires not too much conscious or reflective thought. The lifeworld is what has shaped them. It is what has made them who they are. It is what they like and unreflectively dislike. It is who they are. The underlying attributes of lifeworld difference form the basis of identity and subjectivity. These attributes are the fundamental bases of a learner's sense of belonging in an everyday or formal learning setting and in their levels of engagement (Cope and Kalantzis, 2000; Husserl, 1970).

From the point of view of these underlying differences, the gross demographics of second order differences are as often deeply deceptive as they are immediately helpful. Look at the differences between girls and boys within a particular ethnically defined group or within different age groupings. It is not long before the internal differences in lifeworld attributes between members of that group become so obvious as to indicate that the ethnic or gender descriptor is too simple a variable. Second order differences may be a powerfully revealing introduction to learner differences. First order or lifeworld differences, however, are where the realities of difference truly lie.

The lifeworld is bigger and deeper than any agglomeration of items on the list of second order differences. It encapsulates the full spectrum of possible differences across all the students in the classroom. It encompasses the broad dynamics of power and privilege, of history and location, of the accident of birth and life experience. Together, these become critical determiners of educational and social outcomes.

The lifeworld is deeply permeated by difference; in fact, we live in a myriad of diverging and interacting lifeworlds. The individual is uniquely formed at the intersection of many modes of subjectivity; they are a unique concatenation of many group identities, and they live in and through multiple or multilayered identities. Even when considered from the point of view of second order differences, every individual embodies a unique mix of gender, age, ethnicity/race, locale, socioeconomic, and (dis)ability dimensions, and within any one of these dimensions, quite specific and often complex and multiple configurations emerge (for an ethnicity example, the Italian-Australian who has one Jewish grandparent and who speaks limited Italian with two grandparents). An individual partially shares gross demographics and underlying attributes with a wide and overlapping range of groups, but the particular mix of group attributes is invariably unique (Cope and Kalantzis, 1997).

From the point of view of first order differences, however, our fundamental question is: what are the constitutive components of a person's experience and life history, where the social is instantiated in the personal, group attributes in the individual, broad historical forces in the micro dynamics of personal agency and interpersonal interaction? Here we encounter the raw materials of difference—including human experiences, dispositions, sensibilities, epistemologies, and world views. These are always far more varied and complex than the immediate sight of the demographics would suggest.

Learning succeeds or fails to the extent that it engages the varied subjectivities of learners. Engagement produces opportunity, equity, and participation. Failure to engage produces failure, disadvantage, and inequality. The dilemma for teaching is that, no matter how much filtering is done according to the second order categories (by age, or "ability," or social destiny, or gender, or ethnicity, for instance), groups of learners invariably remain different. For behind the demographics are real people, people who have always-already learned and whose range of learning possibilities are both boundless and circumscribed by what they have learned already and what they have become through that learning.

Education, then, needs to start with a recognition of difference far deeper than the gross demographics of second order differences would allow. The challenge, then, is how do we engage all learners in classrooms of deep difference? In other words, how do we do diversity? How do we negotiate divergence?

Diversity in Action

Diversity has been a factor in modern, mass-institutionalized education from its beginnings. Only recently has it been named as such. Modernity in general and modern education in particular has been marked by three major paradigms for the negotiation of differences. We will name these "separation," "recognition," and "inclusion."

Diversity Dynamics

Separation
Recognition
Inclusion

Listing these as a threesome, however, elides the fact that they exist on barely comparable planes of social action. Separation was often not a conscious agenda; rather, it is a set of practices of avoidance, of dealing with difference by avoiding or refusing to deal with difference. Recognition deals with difference, to be sure, but in a limited and at times tortured way, grounded in the conceptual and practical difficulties of demographic categorization. Inclusion is more a project than a social reality, a series of mostly unrealized agendas, strategic possibilities, and emergent practices rather than a concrete social reality that can be shown to exist to any significant degree.

Separation

The modern institution of compulsory mass education has always had to address deep differences amongst its learners. It does this initially by processes of separation, by means of which a dominant social group maintains its sameness. One of two mechanisms is used: exclusion or assimilation.

Exclusion is a process of refusing to allow people into a social group who are different in defined ways. The exclusionary school, for instance, does not allow students who are the wrong race to attend; the single-sex school does not allow in students who are of the other sex; the wealthy school (consequentially even when not as a matter of principle) excludes students whose families can't afford to pay the fees. Today, some forms of exclusion have been made illegal or are regarded to be repugnant—apartheid, discrimination, racism, or sexism, for instance.

The agenda of exclusion has at times been institutional—when certain types of persons are sent to certain schools or put in certain classes at schools. At other times it has been ideological, when certain narratives, explanations, ways of speaking, and ways of seeing the world are simply left out or even explicitly condemned.

A second mechanism of separation is assimilation. Here, the assimilating social group is just as concerned as the separatist one to make virtue of homogeneous community, but it does not use the same logic or methods. This group says something like, "we will accept people who are different, so long as they become like us." The assimilating school, for instance, makes little allowance for the lifeworld differences amongst its learners. It simply immerses everybody in the singular culture and curriculum of the school. If that's what's good for anybody, then that's what everybody is going to get. Whether you feel comfortable or alienated in this environment, or whether you succeed or fail, is up to you, the learner. You may choose to rise to the "values" and "standards" of the school, and if you don't, you are likely to fail—then have nobody to blame but yourself.

If and when an education system holds out hope of transforming differences in the direction of the norm—the process of assimilation—it may decide to do this as a conscious program. This may be based on a deficit model according to which variations from the norm require treatment or remediation.

Alternatively, there need be no conscious program or agenda. Assimilation in this case rests on an assumption or a hope that students will swim (assimilate themselves) rather than sink (leading to a form of exclusion where, conveniently for the system, the learner can blame themselves). When a learner sinks, it is frequently accompanied by resistance that is experienced as a discipline problem by the school, or it is naturalized by calling it lack of "ability," or it is branded as laziness on the assumption that the student could have risen to the occasion if only the student had tried hard enough.

Education as assimilation means getting into institutionalized education and succeeding at it by crossing over and by making yourself over in the process. You have to leave behind the old self of your lifeworld as hitherto experienced and get on with the strength of those lifeworlds closest to the culture of education. For some, this is a short and comfortable journey. For others, it is long and painful.

Recognition

Since the mid-twentieth century, particularly, ideologies and practices of separatism have been thrown into question, and with this their assumption that the members of a well-functioning social group should be more or less the same. Many of the most distinct forms of strict institutional separation have been challenged—racial segregation or apartheid, the strict delineation or men's and women's jobs and roles, and the institutional separation of disabled people from able-bodied people, for instance. The discourses and practices of sexism, racism, homophobia, and discrimination have come to be regarded as unacceptable. Universal covenants have been developed that create a conceptual and international-legal basis for human rights. These changes are the result, in part, of political contestation and institutional change. They are also a consequence of changing social conditions and shifts in values.

At first, the shift away from separatism and assimilation is a cautious one, a step in the direction of the recognition of differences. Instead of pretending that the differences don't exist, or hoping they will go away, or trying to make them go away, differences are recognized and at least minimal measures are taken to try to produce more equitable outcomes.

Sometimes, the process of recognition is no more than that. People come to recognize publicly that there are big differences in the ways people live and speak and think in their everyday lifeworlds. But they may go little further than this and say, as French speakers would, "laissez-faire," or let things be the way they are. There's nothing much we need to do. If differences have a life of their own, they are best left alone. "I'm happy with my way of life," a proponent of laissez-faire might say, "and I'm happy that you seem to be happy with yours. So let's leave things the way they are. I won't interfere in your lifeworld if you don't interfere in mine. If I have a private opinion about your way of life, I'm going to keep it to myself."

Such a view of diversity is primarily descriptive. It recognizes differences and leaves them alone, except when they demand attention. A certain kind of limited recognition of second order or demographic difference does no more than reflect diversity in the world as it is found. It is as if differences were to be encountered like fragile heritage objects in a museum, to be handled with care to ensure that they are maintained, to be labeled with static and essentialized demographic categories.

Furthermore, if the politics of recognition is to take one further step, becoming more interventionary, its agenda is to right historic injustices by developing special programs for people who fall into broad demographic categories: socioeconomic status, race, ethnicity, gender, and the like. If you need to do something to improve people's conditions of life, you build special programs for each group, such as for the poor, for women, for blacks, or for indigenous peoples.

Both of these modes of recognition—laissez-faire and reformist—have their limitations. The laissez-faire approach does little or nothing to right historic injustices. To free people from the direct impact of legal, structural, or attitudinal racism or sexism may allow them to succeed in ways that were not previously possible. But if the conditions are not favorable, if the support mechanisms are not there that will enable them to succeed—the material resources or the cultural capital, for instance—nothing will change. And when nothing changes, it looks as though inequality and poverty is the fault of its victims, not the social system or environment.

As important as recognition might be as a mark of acknowledgement and a strategy to make students and families feel they belong, however, it does not necessarily tackle the question of access, or how education might provide minority students enhanced access to avenues of social power and material resources (Delpit, 1988). Sometimes, this kind of recognition of difference feels tokenistic or patronizing even. Other times it creates stereotypes, where a child has to go searching for the school's vision of what is sufficiently exotic to be different. At still other times, it creates categories that are all-too-neat and which oversimplify the more complex realities of identity in ways that are less than helpful.

Inclusion

So, how does one move beyond these limited forms of recognition? What actually makes them different? The answer, in part, is to work with the deeper, more complex, multilayered and overlapping, and ever-widening differences between lifeworlds—life experiences, interests, orientations, values, stances, dispositions, sensibilities, communication styles, interpersonal manners, ways of thinking, and preferred approaches to learning. These underlying lifeworld attributes are what really make boys and girls different, and people of one ethnic group different from another—but not always in ways that fit conventional or expected patterns. The categories of difference delineated in the gross demographics all come into play, but often not in predictable ways. In fact, one of the reasons the demographics oversimplify things is the multilayered nature of identity, where, for any one person, the various aspects of their life come together in an always-unique mix. It would be too easy, however, just to say "we're all individuals, so let's not worry about diversity." On the contrary, the demographics tell of group histories and experiences that do come to play for individuals, whether they seem to conform to the group norm or to represent an exception. These group-related aspects of difference are always falling together for any one person in ways that are unique and never clearly predictable from demographic profiling.

How, then, do we work with a reading of diversity that is based on lifeworld attributes—the way a person feels, identifies, speaks, thinks, and acts? Rather than neatly categorized groups, we need to think of difference as a series of layers, intersections, and matrices. We need to be aware of broad variations within groups that would otherwise seem to be easily open to categorization. Groups need to be viewed as clearly distinguishable in some respects, but also in various kinds of relationship or tension in which the one in a group defines the other—masculine and feminine, rich and poor, in-group and out-group.

Difference, moreover, needs to be viewed as an active and dynamic state, a kind of begging to differ, the result of which is that the lifeworld is constantly being transformed by its own inhabitants. In an earlier modernity, the trend was to homogenization. Today, the trend is towards divergence. Divergence goes one step beyond diversity as a state to recognize the active, dynamic flux that is the result of both difference and diversity. Along the way, groups may create moments of strategic autonomy—a women's support group or an ethnic school, for instance—but the agenda of the politics of divergence is as much as anything to create a more open and inclusive mainstream, be that in domains such as employment, democratic citizenship, or schools.

Nor are the differences simply to be accepted. Some aspects of difference are regarded by at least some of their subjects as less than satisfactory and open to contestation, such as social and cultural practices that contribute to material inequality, or racism, or sexism. Whereas the concept of diversity may remain impassively descriptive, an understanding of the dynamics of divergence could entail an active change agenda in the interests of social innovation.

Divergence, by contrast with more static notions of the recognition of diversity, works on agendas for action. It is inclusive, bringing differences together and working to rebuild the world in a collaborative way. It works on the productive tensions between differences, rather than shrugging one's shoulders and saying "that's life," the way things will always be destined to be.

A more inclusive education for diversity will work with learners to invent and reinvent themselves and their worlds. Such an approach may be driven by a range of social and self transformative agendas, from establishing and building a career, to entrepreneurial innovation, to ethical concern for social justice or the environment. Such an inclusive education addresses the active processes of divergence and inclusion across a full spectrum of potential individual and social objectives, from the pragmatics of self advancement to the social ethical agendas of emancipation and sustainability.

Inclusion is a way of working such that differences are without prejudice to social access and symbolic recognition. Inclusive education means that you don't have to be the same to have similar opportunities: not identical opportunities, but the same kinds of opportunities measured in terms of comparable access to material resources through employment, political participation, and senses of belonging to a broader as well as a localized community. Inclusion involves a subtle but profound shift from a more superficial multiculturalism of recognition. It means that the mainstream—be that the culture of the dominant groups or institutional structures such as education—is itself transformed. Instead of representing a single cultural destination, the mainstream is a site of openness, negotiation, experimentation, and the interrelation of alternative frameworks and mindsets.

Learning, then, is not a matter of "development" in which you leave your old selves behind, leaving behind lifeworlds that would otherwise have been framed by education as more or less inadequate to the task of modern life. Rather, learning is a matter of repertoire, starting with recognition of lifeworld experience and using that experience as a basis for extending what one knows and what one can do. The pluralist process of transformation, then, is not a matter of vertical progress but one of expanding horizons. These new horizons do have an impact on the lifeworld: learners still engage in and with their lifeworlds in new ways, but not necessarily in order to leave those lifeworlds behind in a kind of one way trip.

References

Cope, B. and Kalantzis, M. (1997). *Productive Diversity: A New Approach to Work and Management*. Sydney: Pluto Press.

———. (2000). "Designs for Social Futures." In B. Cope and M. Kalantzis (eds.), *Multiliteracies: Literacy Learning and the Design of Social Futures*. London: Routledge. 203–234.

Delpit, L. D. (1988). "The Silenced Dialogue: Power and Pedagogy in Educating Other People's Children." *Harvard Educational Review*, 58, 280–298.

Department of Immigration and Multicultural Affairs. (1998). "Charter of Public Service in a Culturally Diverse Society." Canberra: DIMA.

Husserl, E. (1970). *The Crisis of European Sciences and Transcendental Phenomenology*. Evanston: Northwestern University Press.

Kalantzis, M. and Cope, B. (2006). "On Globalisation and Diversity." *Computers and Composition*, 31, 402–411.

———. (2008). *New Learning: Elements of a Science of Education*. Cambridge: Cambridge University Press.

CHAPTER 2

Language as Racism:
A New Policy of Exclusion

Panayota Gounari and Donaldo Macedo

*In the first place we should insist if the immigrant who comes
here in good faith becomes an American and assimilates himself to us, he shall be treated with an exact equality with everyone else, for it is an outrage to discriminate against such man
because of creed, or birth place, or origin…But this is predicated
upon the man's becoming in very fact an American and nothing
but American…There can be no divided allegiance here. Any
man who says he is an American but something else also, isn't an
American at all…We have room for but one language in this
country and that is the English language…and we have room for
but one loyalty and that is a loyalty to the American people.*

—*Theodore Roosevelt, 1907*

As Dinesh D'Souza and other cultural commissars falsely proclaim
the end of racism, language has become the last refuge to wantonly
discriminate with impunity. Often these cultural commissars rely
on selective history, as Lou Dobbs, a CNN commentator, astutely
quoted Theodore Roosevelt to make the case that to be an American
"does not allow for divided loyalties" and that speaking a language
other than English is tantamount to disloyalty. Leaving aside the
blatant contradictions inherent in Roosevelt's definition of what it
means to be an American, what is never interrogated is the undemocratic proposition stating that in order to be, one must stop being.
That is, to be an American requires that immigrants commit both

cultural and linguistic suicide since, as Roosevelt insisted, "there is no room in this country for hyphenated Americanism" (Crawford, 2004, p. 67). In essence, these cultural commissars fail to acknowledge that the requirement to blindly assimilate so that one can become an American represents also a quasi cultural genocide that is designed to enable the dominant cultural group to consolidate its cultural and linguistic hegemony. As correctly pointed out by Amilcar Cabral, the ideal for cultural domination can be reduced to the following: The dominant cultural group (1) liquidates practically all the population of the dominated country, thereby eliminating the possibilities for cultural resistance; or (2) succeeds in imposing itself without damage to the culture of the dominated people—that is, harmonizes economic and political domination of these people with their cultural personality (Cabral, 1973, p. 40).

The first strategy was used to a great extent by European colonizers in their quasi-genocide of Native Americans. The second strategy provides the basis for the violent assimilation policies, which differ little from the "imperialist colonial domination [that] tried to create theories which, in fact, are only gross formulations of racism, and which, in practice, are translated into a permanent state of siege on the indigenous populations" (Cabral, 1973, p. 40).

In other words, the "melting pot" theory that has been recycled since the beginning of the century and is now used again by the proponents of English-only, or the "progressive assimilation of the native populations[,]...turns out to be only a more or less violent attempt to deny the culture [and language] of the people in question" (Cabral, 1973, p. 40). It is this racist denial of cultural and linguistic rights that needs to be fully understood in the current language policy debates, instead of allowing it to obfuscate the binaristic position of Americanism versus multiculturalism. The real issue in the language debate is cultural and economic domination and racism. In fact, it is an oxymoron to speak of American democracy and "our common culture" in view of the quasi-apartheid conditions that have predominated in the United States. It is precisely because of the power inherent in language and culture that even liberal scholars such as Arthur Schlesinger, Jr. became concerned that a "cult of ethnicity has arisen both among non-Anglo whites and among nonwhite minorities to denounce the idea of a melting pot, to challenge the concept of 'one people' and to protect, promote, and perpetuate separate ethnic and racial communities" (Schlesinger Jr., 1992, p. 15). Schlesinger's position not only was dishonest but also served to alarm the public regarding what he referred to as the "multiethnic dogma [that] abandons historic purposes, replacing assimilation by fragmentation, integration by separatism. It belittles Unum, and glorifies Pluribus" (Schlesinger Jr., 1992, p. 15). A more honest account of history would highlight the fact that African-Americans did not create laws so they could be enslaved; they did not promulgate legislation that made it a crime for them to be educated; nor did they create redlining policies that sentenced them to ghettos and segregated schools and neighborhoods. Unless Arthur Schlesinger was willing to confront the historical truth, his concern for the disuniting of America is yet another veil to mask white-male-supremacy values that place the discriminatory policies in the United States beyond analysis—thus, beyond scrutiny. What Schlesinger failed to recognize was that there was never a "common culture" in which people of all races and cultures equally participated. The United States was founded on a cultural and linguistic hegemony that privileged and assigned control to the white patriarchy and relegated other racial, cultural, and gender groups to a culture of silence.

While the cultural commissars neglect to acknowledge the racist and antidemocratic nature of their hegemonic policies that parade under the rubric of democracy and our "common culture," the pain of these violent racist policies was always felt and denounced, as exemplified by the American Indians at Wounded Knee:

> We have not asked you to give up your religions and beliefs for ours.
> We have not asked you to give up your language for ours.
> We have not asked you to give up your ways of life for ours.
> We have not asked you to give up your government for ours.
> Why can you not accord us the same respect? (Lyons, 1973)

These questions represent the essence of what it means to be engaged in bilingual pedagogy, and for us bilingualism represents what it means to experience multiple realities by negotiating two languages, two cultures, and the humane meshing of manifold identities that become layered in who we become. It is within this critical spirit and the simple call for our humanity that we want to situate our discussion of the current language policy in the United States.

In addition to our analysis of multiple referenda designed to tongue-tie immigrant students in their native languages, we also highlight in this chapter the importance of tensions, contradictions, fears, doubts, hopes, and dreams involved in the process of understanding what it means to learn English. These tensions, contradictions, fears, doubts, hopes, and dreams are never captured in the pragmatism and technicism of the so-called objectivity and accountability of standardized tests currently imposed under the *No Child Left Behind* federal mandate. Technicism, however, itself is not technical, it is deeply historical. What becomes clear to most students in their struggles to learn English is that standardized tests never create the necessary pedagogical structures that would enable submerged voices to emerge. What the proponents of these tests fail to realize is that cultural voice can never be reduced to a dehumanized "objective" test score.

Unfortunately, cultural voice is never a concern for the positivistic school of thought where number crunching is always given primacy while human aspirations and dreams are usually relegated to the margins of our education. What is particularly interesting is that positivism is always used in the name of "progress." The influence of positivism with its theological rigidity toward "objectivity" is not felt solely through the imposition of the present misguided high stakes testing—its influence has also reached the reading and literacy fields under the veil of "Back to Basics" through, for example, the militarism of the *Reading First Program*. Behind the raw test scores there is always a human face whose humanity is often sacrificed at the altar of accountability and objectivity. What the proponents of standardized testing in English-only fail to acknowledge is that these tests can never capture the ambivalence of our fractured cultural soul yearning to make meaning out of a bittersweet existence in the undemocratic requirement that in order to be in our democratic society, one must assimilate. That is, one's success is measured by the degree to which one leaves behind one's culture and one's language so they can be frozen in time and space. What proponents of these tests fail to admit is that their scores embellished by the jargon of validity and reliability can never reflect the tensions, contradictions, fears, doubts, hopes, and deferred dreams that are part and parcel of living in a borrowed reality. We say borrowed reality to the degree that no matter how much we attempt to assimilate so as to escape the rap of discrimination, we are always forced to reduce our humanity to clear-cut little boxes as African-Americans, Hispanic-Americans, American-Indians, and worse, as Glória Anzaldúa so painfully reminded us:

> Deslenguadas. Somos los del español deficiente. We are your linguistic nightmare, your linguistic aberration, your linguistic mestisaje, the subject of your burla. Because we speak with tongues of fire we are culturally crucified. Racially, culturally and linguistically. Somos húerfanos—we speak an orphan tongue. (Anzaldúa, 1987, p. 103)

The erasure of one's cultural dignity in *otherness* is best characterized by an existence that is almost culturally schizophrenic—that is, being present and yet not visible, being visible and yet not present. It is a condition that invariably presents itself to the reality of linguistic minority students—the juggling of two worlds, two cultures, and two languages. It is a process unrecognized by test scores and through which we come to know what it means to be at the periphery of the intimate and yet fragile relationship between two cultural worlds where the subordinate cultural being is always required to declare where we come from. We must always fulfill the American cultural need to ethnically typecast so as to devalue other non-American cultural beings. As soon as one opens one's mouth, one is met with the almost automatic "Where are you from?" or worse, "*What* are you?" It is a condition that leads to a pedagogy of entrapment to the degree that it requires of the cultural "other" what the system denies him or her. In other words, as we are asked to assimilate as rapidly as possible into English and the American culture, the same ideology creates obstacles that would make total assimilation impossible.

This is in line with Steward Hall's critique of the British colonialism when he discusses what it means to be a Jamaican in England. By substituting British for American, we can obtain the same reality. Although our linguistic minority students are required to be educated in English-only which could enable some of them to know America from the inside, the sad reality is that they will always end up in the same reality as Steward Hall: "I knew [America] from the inside. But I'm not and never will be [American]. I know both places intimately, but I am not wholly of either place. And that's exactly the [immigrant] experience, far away enough to experience the sense of exile and loss, close enough to understand the enigma of an always-postponed 'arrival'" (Morley and Chen, 1996). This always-postponed "arrival" mirrors Langston Hughes' "deferred dream" of African-Americans whose continued subordination and relegation to ghetto life makes a lie of the English-only propositions that education in English-only will guarantee linguistic minority students success in school and access to the high political and economic echelons.

In order to understand the significance of the lie inherent in these false propositions, we need to ask two fundamental questions: First, if English is the most effective language of instruction, how can we explain that 11 million adults are illiterate and some 30 million have "below basic" skills in prose (Feller, 2005)? Second, if Education in "English only" can guarantee linguistic minority students a better future, as William Bennett and Ron Unz promise, why do the majority of black Americans, whose ancestors have been speaking English for over two hundred years, find themselves still relegated to ghettos? It is the same William Bennett who is now suggesting that "if you wanted to reduce crime, you could abort every black baby in this country, and your crime rate would go down" (Macedo Freire, 2002).

Against a landscape that gives primacy to test scores in English and methodological fads over the crucial understanding of social construction of concepts in the first place, it is not a coincidence that the multiple referenda designed to impose English as the only language of instruction constitute, in reality, an attempt to turn what had been a *de facto* into a *de jure* language policy. True, there was never an official language policy in the United States, but never before, after the civil rights movement, has the legislative narrative been so blatantly exclusionary and racist in terms of languages other than English. Conservatives have made it a "moral" imperative to push for English monolingualism, not only because all other languages have been perceived as threatening the hegemony of English, but also because in the commonsense discourse, knowledge of the English language is equated with the "common good," since, according to the Massachusetts Proposition 2, for instance,

> [T]he government and the public schools of Massachusetts have a moral obligation and a constitutional duty to provide all of Massachusetts's children, regardless of their ethnicity or national origins, with the skills necessary to become productive members of our society. Of these skills, literacy in the English language is among the most important. [1.1 (c), 2001] (An Initiative, 2001)

Within this framework, the anti-bilingual education initiative owed a large part of its success to the use and dissemination of a commonsense discourse about what constitutes the "common good." Through this discourse, the proposition for abolishing bilingual education programs for linguistic minority students appeared as striving for a common good (that is, proficiency in English). In this respect, the anti-bilingual education initiative positioned English as education in and of itself and used a language of morality and "equality" to justify the need for monolingual classrooms, blatantly rejecting and ignoring years and thousands of pages of research proving the exact opposite. By putting in the forefront of their campaign the concern for linguistic minority students' success and their participation in the American mainstream, opponents of bilingual education managed to manipulate the public discourse about the "common good."

The overcelebration of English as a common global language gains more importance in light of the nonexistent official language policy in the United States. The hegemony of English as a lingua franca is presented as a natural and given fact that allows no space for questioning. In this way, commonsense discourse works to manipulate public opinion into embracing an idea of a common language that necessarily excludes from the debate all the "uncommon languages." The construction of "commonality" as exclusion finds its justification in the discourse of common sense. Common sense needs to be understood in conjunction with the thinking of Antonio Gramsci. It is "the conception of the world which is uncritically absorbed by the various social and cultural environments in which the moral individuality of the average man is developed" (Gramsci, 1971). He insists that common sense is not a single unique conception that is identical in time and space, in that it evolves and changes in each historical/social/cultural or geographical locus. Its most fundamental characteristic is that "it is a conception which, even in the brain of one individual, is fragmentary, incoherent and inconsequential, in conformity with the social and cultural position of those masses whose philosophy it is" (Gramsci, 1971, p. 343).

One of the reasons the language of common sense appears to be natural and true is that it has been dehistoricized. While discourses are necessarily historical since they are constructed and shaped in different spatiotemporal contexts, the current debate on bilingual education in the United States conveniently leaves out the inextricable relation between language and culture. As a result, it erases a long history of immigration, bilingualism, linguistic oppression, and racism and reduces the issue to simply "teaching English." While this discourse closes down the discussion and analysis vital to the continued existence of any democracy, it also robs people of any opportunity to shed light on their personal civic role to such a degree that they embrace it with the utmost faithfulness and respect, as an absolute "fact of nature." If common sense is the assimilation of the dominant ideology to the degree that it seems natural and is uncritically believed, then the discourse of common sense used by the dominant order can be understood as the uses of language as a form of social practice that works to neutralize language and therefore the ensuing practices, institutions, assumptions, and presuppositions. All this is shaped through historical, social, cultural, and ideological practices that, in the case of common sense, are either erased or invisible, making the discourse of common sense a powerful tool to justify policies, political decisions, and practices that are largely designated to oppress, stupidify, and block dissent. Therefore, people not only embrace this commonsensical discourse but, at the same time, they create, recreate, and redefine it. Along these lines, the selection of a specific discourse around the worth of languages other than English does not allow any possibility for interrogation, which might lead to the opening up of debate. In other words, instead of viewing the increasing influx of immigrants and the challenges this poses to education as containing the possibility for developing multiple referents for understanding, developing and implementing more inclusive curricula, valorizing the students' identities, and increasing their access to cultural and economic goods, conservative scholars and policy-makers recoil into a fixed monolingual space where the existence of any other language presents a permanent threat to standard English. In the dominant public discourse the assumption is that English is our "common language" and that it is being threatened. Samuel Huntington is blunt in his apocalyptic rhetoric. He claims that "immigrants, especially those from Mexico, are undermining the 'Anglo-Protestant creed,' destroying the shared identity that makes us Americans. These immigrants do so by refusing to assimilate, to learn English, and to become American citizens and by maintaining a segregated society centered on un-American values." Huntington insists that, if Mexican-Americans learn English but maintain Spanish as their second language, it is an indication that they are refusing to become good Americans (Etzioni, 2004, p. 48).

Within this framework, language cannot be seen as only a neutral tool for communication. It must be viewed as the only means through which learners make sense of their world and transform it in the process of meaning-making. In the meaning-making process, both subordinate students and their teachers need to know that standard English is "the oppressor's language yet I need it to talk to you" (hooks, 1994, p. 168). As bell hooks so painfully understands, standard English "is the language of conquest and domination . . . it is the mask which hides the loss of so many tongues, all those sounds of diverse, native communities we will never hear" (hooks, 1994, p. 168).

In the following sections we will discuss how language policy in the United States, which has largely promoted monolingualism, has given rise to different English-only movements across America. We will also attempt to analyze the hidden agenda behind the anti-bilingual education proponents' discourse.

LANGUAGE POLICY AND ENGLISH-ONLY MOVEMENTS IN THE UNITED STATES

It often comes as a surprise that in the United States there is no official language policy. The commonsense assumption is that English is the official language, although this has never been officially legislated. In this respect, by consistently avoiding legislating an official language policy, regulated by legal and constitutional declaration, the United States has been the envy of many nations that aggressively reinforce language use within their borders through explicit policies designed to protect the "purity" and "integrity" of the national language. Even without a rigid policy, the United States has managed to achieve such a high level of monolingualism that speaking a language other than English constitutes a real liability. American monolingualism is part and parcel of an assimilationist ideology that decimated the American indigenous languages as well as the many languages brought to this shore by various waves of immigrants. As the mainstream culture felt threatened by the presence of multiple languages, which were perceived as competing with English, the reaction by the media, educational institutions, and government agencies was to launch periodic assaults on languages other than English. This was the case with American-Indian languages during the colonial period and German during the two world wars.

This covert assimilationist policy in the United States has been so successful in the creation of an ever-increasing linguistic xenophobia that most educators, including critical educators, have either blindly embraced the dominant assimilationist ideology or have remained ambivalent with respect to the worth of languages other than English. The assumption that English is a more viable and pedagogically suitable language than others has completely permeated U.S. educational discourse. Opponents of bilingual education, conservative educators, and advocates of movements that support national and linguistic homogeneity and assimilation, assign to language a mechanistic, technical character. Within this technical perspective, they propose the adoption of English-only instruction as a remedy for the so-called "failure" of linguistic minority students. However, what they fail to see is that language as a social practice shapes human existence in a dual way. For one, it affects the way humans are perceived through their language. Second, individuals develop *discourses*, that is, systems of communication shaped through historical, social, cultural, and ideological practices, which can work to either confirm or deny the life histories and experiences of the people who use them. Therefore, the claim that English fluency as a simple skill will come in monolingual classrooms is very much part of a commonsense discourse that, under the pretence of the "common good," veils inevitable exclusions. In this respect, this proposition is hegemonic in that a group of people has been able to define what the common good is and it has imposed this monolithic view by privileging some meanings over others (Mouffe, 1999).

Another issue worth analyzing in the anti-bilingual education initiative is the promise of a better life that will come from becoming fluent in English. According to the same petition, "[I]mmigrant parents are eager to have their children become fluent and literate in English, thereby allowing them to fully participate in the American Dream of economic and social advancement." Through the proposition that the English language is a passport that gives access to the higher cultural, political, and economic echelons of U.S. society, opponents of bilingual education attempt to hide their ongoing cultural invasion of other groups. Learning standard English will not iron out social stratification, racism, and xenophobia. Nevertheless, under the "naivete" pretext and the notion that language exists in a vacuum, conservative educators continue to disarticulate language from its social and ideological context by conveniently ignoring the facts discussed below.

The real context of the debate has nothing to do with language itself, but with what language carries in terms of cultural and economic goods. As Bhabha reminds us, the problem of cultural difference is produced when

> there is some particular issue about the redistribution of goods between cultures, or the funding of cultures, or the emergence of minorities or immigrants in a situation of resources—where resource allocation has to go—or the construction of schools and the decision about whether the school should be bilingual or trilingual or whatever. (Bhabha, 1999, p. 16)

If Bhabha is correct, then linguists, educators, and policy-makers need to move beyond the notion that language is a "treasure," a common possession—what Bourdieu called "the illusion of linguistic communism" (Bourdieu, 1999). The existence of a common language, a "code" open to use by everybody and equally accessible to all—as assumed by proponents of the English-only movement—is illusionary. This assumption begs the question of why, from a sea of languages, "dialects," "standards," and "varieties," standard English emerged as the most appropriate and viable tool of institutional communication. Application of the simple theorem that "language is identified with its speakers" would require that we find native speakers of standard English, identify them, and analyze their "mother tongue." I believe it would be fair to say that no American is a native speaker of "Harvard English," and definitely no French person has the discourse of the Académie Française as his or her "mother tongue." If mastery of standard English is a prerequisite for enjoying the "common culture," we first need to clarify what kind of standard English we are to teach and thus to speak.

This statement seems to contradict itself, as some would argue that there are not many kinds of standard English. Standard English would literally be "clear" English, sterilized from any "familiarity," "jargons," or "unacceptable" forms that "dialects" often use, the kind of English used in the "Great Books." In addition, the existence and use of a "colorless and odorless" sterilized code implies that language is dehistoricized and that we, as humans, have no obvious markers of identity (such as ethnicity, culture, race, class, gender, or sexual orientation) reflected and refracted through our language. A more honest definition would address the following questions: "Who speaks the standard?" "Who has access to it?" "Where does one develop this particular discourse and through what process does one access apprenticeship in a particular discourse?"

In this context it is not an exaggeration to speak about linguistic hegemony to the extent that the development of a normative discourse through standard English naturalizes, for instance, ideologies and practices that are connected to white supremacy, racism, and oppression. According to Fairclough, "naturalized discourse conventions are a most effective mechanism for sustaining and reproducing cultural and ideological dimensions of hegemony" (Fairclough, 1995, p. 7).

As we have suggested, the existence of a common language also implies the existence of a common culture. Conversely, any reference to a common culture must also imply the existence of an uncommon culture. What supporters of the English-only movement and opponents of bilingual education wish to achieve through the imposition of a "common culture" is the creation of a de facto silent majority. Since language is so intertwined with culture, any call for a "common culture" must invariably require the existence of a "common language."

In fact, the English-only proponents' imposition of standard English as the only viable vehicle of communication in our society's institutional and civic life, under the rubric of our "common language," inevitably leads to the "tongue-tying of America." This "tongue-tying" aids the conservative attempt to reproduce dominant cultural values by insisting, on one hand, on ever-present, collective myths that present a diverse origin, a diverse past, and diverse ancestors, and, on the other, on a common mother tongue and a necessary common, homogeneous, and indivisible future (Memmi, 1996). In general, movements that claim to promote ethnic, linguistic, and cultural integrity attempt, in reality, to impose cultural domination through linguistic domination under the guise of an assimilative and let's-live-all-together-happily model. This process invariably becomes a form of stealing one's language, which is like stealing one's history, one's culture, one's own life. As Ngugi Wa'Thiongo so clearly points out:

> Communication between human beings is the basis and process of evolving culture. Values are the basis of people's identity, their sense of particularity as members of the human race. All this is carried by language. Language as culture is the collective memory bank of a people's experience in history. Culture is almost indistinguishable from the language that makes possible its genesis, growth, banking, articulation, and indeed its transmission from one generation to the next. (Wa'Thiongo, 1996)

Hence, the proposition "common culture" is a euphemism that has been used to describe the imposition of Western dominant culture in order to eliminate, degrade, and devalue anything different. It is a process through which the dominant social groups attempt to achieve cultural hegemony by imposing a mythical "common language." In turn, language is often used by the dominant groups as a manipulative tool to achieve hegemonic control. As a result, the current debate over bilingual education has very little to do with language per se. The real issue that undergirds the English-only movements is the economic, social, and political control of a largely subordinate majority by a dominant minority, which no longer fits the profile of what it means to be part of "our common culture" and to speak "our common language."

The English-only movements' call for a "common language" does more than hide a pernicious social and cultural agenda. It is also part of an attempt to reorganize a "cultural hegemony," as evidenced by the unrelenting attack of conservative educators on multicultural education and curriculum diversity. The assault by conservatives on the multiplicity of languages spoken in the United States is part of the dominant cultural agenda to both promote a monolithic ideology and to eradicate any and all forms of cultural expression that do not conform to the promoted monolithic ideology.

What becomes clear in our discussion so far, is that the current bilingual education debate has very little to do with teaching or not teaching English to non-English-speakers. The real issue has a great deal more to do with the hegemonic forces that aggressively want to maintain the present asymmetry in the distribution of cultural and economic goods.

The Question of Language: Opening up Democratic Spaces

The current debate about whether English is the most viable medium of instruction and the ongoing attack on bilingual education programs points to American society's growing intolerance of the "other." The inability to deconstruct the conservative commonsense discourse further signals the growing indifference for the affairs of the *polis* that are now left in the hands of lobbyists and interest groups. People stop asking questions about the nature of the system that is imposed on them because they begin to perceive such imposition as common sense, a fact of nature. The attack on bilingual education further denies immigrant children a basic human and civil right, namely, the right to learn in their native language. According to Article 29 of the United Nations Convention on the Rights of a Child (1989), "the education of the child should be directed to [...] the development of respect for the child's parents, his or her own cultural identity, language and values, for the national values of the country in which the child is living, the country from which he or she may originate, and for civilizations different from his or her own."

Along the same lines, Article 30 states that "a child belonging to an [ethnic, religious, or linguistic minority] should not be denied the right [...] to use his or her own language." Access to education in one's native language should be intimately connected with the question of democratic practices. No doubt, immigrant learners need to learn the language of the host country, but this should happen in a way that will enable them to not only read the *word* but also the *world*. It is the oppressor's language, as bell hooks suggests, a language of conquest and domination, a weapon that can shame, humiliate, and colonize, silence and censor, yet "I need it to talk to you" (hooks, 1994, p. 168). In this respect, how will the power of the word redefine the power in the world? How does one redefine and/or reinvent the oppressor's language? How do we make it a counter-hegemonic discourse? The reappropriation of the language of the oppressor together with preserving our native tongues should come along with the redistribution of wealth and power in the U.S. society. This means equal participation of immigrants to the U.S. society, educational opportunities and resources, a sense of citizenship that they belong here as well, opportunities for involvement in public affairs, and representation in the government and other sectors of public life. It also means that the teaching of English should come with an understanding of what the hegemonic language carries with it and how it can be used to oppress and empower at the same time. Accordingly, the educational dimension of English has to bear relevance to students' lives and histories today. Learning English should move beyond the teaching of grammar and language skills, to encourage students to raise critical questions about their own social, cultural, and historical location. In this sense, English will work in more directions than simply translating meanings from one language to the other. It will enable students as members of traditionally oppressed and marginalized groups to translate their private troubles into public issues. No individual or social, cultural, or ethnic groups can start the struggle for self-affirmation without the use of their native language (Freire, 1995). For oppressed and marginalized groups the need for self-determination is crucial in the shaping and reshaping of their identities as they struggle to negotiate the new realities of the host country and to position themselves in the distribution of cultural and economic goods.

At the same time, public discourse around the issue of educating linguistic minority students should switch terrain to include questions about access to cultural, social, and economic goods, language hierarchies, ideology, and power. The consensus around language that English-only proponents profess is hypocritical and undermines the very foundations of democracy. Beyond the important issue of denying linguistic minority students their human rights, the commonsense discourse perpetuates economic and social inequalities. The "common language" argument necessarily situates many other languages in the periphery and obviously fails to question the hegemony of English worldwide as well as the privileges of those who have access to it. This happens largely because conservative ideologies have been legitimized and disseminated through a well-organized network of diffusion that has succeeded in presenting monolingualism as the inevitable "common good." Consequently, there is a vital need to break the continuity and consensus of common sense that currently dominates the language debate in the United States. This is particularly important given that the struggle takes place in schools that are deeply political spaces. In these pedagogical spaces students should be able to understand how power works within educational institutions to legitimize some languages and some forms of knowledge, namely, Westernized knowledge, at the expense of other subjugated knowledges, histories, identities, and discourses. They will be able to treat knowledge as a contested field and as part of a project of politics and emancipation.

Ultimately, as we have attempted to demonstrate, the question of language is a deeply political one and it should always be understood in relation to economic, social, and cultural hierarchies. June Jordan's comments are a propos here:

> I am talking about majority problems of language in a democratic state, problems of a currency that someone has stolen and hidden away and then homogenized into an official "English" language that can only express non-events involving nobody responsible, or lies. If we lived in a democratic state our language would have to hurtle, fly, curse, and sing, in all the common American names, all the undeniable and representative participating voices of everybody here. We would not tolerate the language of the powerful and, thereby, lose all respect for words, per se. We would make our language conform to the truth of our many selves and we would make our language lead us into the equality of power that a democratic state must represent. (hooks, 1994, p. 168)

NOTE

Some parts of this chapter were published elsewhere in earlier publications.

REFERENCES

An Initiative Petition for a Law: An Act Relative to the Teaching of English in the Public Schools. 01–11, Boston (MA), July 31, 2001.

Anzaldúa, G. (1987). *Borderlands: The New Mestiza*. San Francisco: Spinster/ Aunt Lute.

Bhabha, H. (1999). "Staging the Politics of Difference: Homi Bhabha's Critical Literacy." In Olson, G. and L. Worsham (eds.), *Race, Rhetoric and the Postcolonial*. New York: State University of New York Press, 165–204.

Bourdieu, P. (1991). *Language and symbolic power*. Cambridge, MA: Harvard University Press.

Cabral, A. (1973). *Return to the Source: Selected Speeches of Amilcar Cabral 9*. New York: Monthly Review Press.

Crawford, J. (2004 [1989]). *Educating English learners: Language Diversity in the Classroom*. 5th ed. Los Angeles: Bilingual Education Services.

Etzioni, A. (2004). "The Real Threat: An Essay on Samuel Huntington." *Contemporary Sociology*, 34(5), 477–485.

Fairclough, N. (1995). *Critical Discourse Analysis: The Critical Study of Language*. New York: Addison Wesley Longman Inc.

Feller, B. (2005). *Study: 11 M U.S. Adults Can't Read English*, Associated Press. Retrieved from http://ro-a.redorbit.com/news/science/331665/study_11_million_us_adults_cant_read_english/index.html.

Freire, P. (1985). *The Politics of Education: Culture, Power and Liberation.* New York: Bergin and Garvey.

Gramsci, A. (1971). *Selections from Prison Notebooks* (ed. and trans. Q. Hoare and G. Smith). New York: International Publishers.

hooks, b. (1994). *Teaching to Transgress: Education as the Practice of Freedom.* New York: Routledge.

Lyons, O. [Onondaga] (1973). *Presenting the Statement of the Grand Council of the Iroquois to the United States Government.*

Macedo, D. and Freire, P. (2002). *Ideology Matters.* Boulder, Colorado: Rowman and Littlefield Publishers.

Memmi, A. (1996). "La patrie littéraire du colonisé." *Le Monde Diplomatique.* September issue.

Morley, D. and Chen, K. (eds.) (1996). *Stuart Hall: Critical Dialogues in Cultural Studies.* New York: Routledge.

Mouffe, C. (1999). "Rethinking Political Community: Chantal Mouffe's Liberal Socialism." In Olson, G. and L. Worsham (eds.) *Race, Rhetoric and the Postcolonial.* New York: State University of New York Press, 165–204.

Schlesinger Jr., A. (1992). *The Disuniting of America: Reflections on a Multicultural Society.* New York: Norton.

Wa'Thiongo, N. (1986). *Decolonizing the Mind: The Politics of Language in African literature.* Portsmouth: Heinemann.

CHAPTER 3

Policies as Top-Down Structures versus Lived Realities: An Investigation of Literacy Policies in Greek Schools

Triantafillia Kostouli and Soula Mitakidou

CULTURES IN INTERACTION: CO-CONSTRUCTING LOCAL CULTURES OF MEANING

This chapter reviews the various kinds of initiatives that have been designed in the Greek context to foster minority students' empowerment, which is traditionally seen as involving students' appropriation of school literacies. Two kinds of programs are surveyed: (a) programs specially designed for students (such as Rom and children from the Muslim religious minority) who use a language other than Greek to communicate in most contexts of their lives; and (b) initiatives operating in mainstream contexts, that is, in classrooms where native and nonnative speakers with varying knowledge of Greek interact.

Policy work in the Greek context has been mainly concerned with "top-down empowerment," assumed to be acquired in a uniform fashion across schools through the implementation of programs designed by experts in the field, though these experts are outsiders to the communities they study. Despite their differences in orientation and design principles, bilingual programs existing in the Greek elementary school system seem to build upon a rather static notion of student empowerment. This is seen as a top-down process, contingent upon teacher adherence to their policies that privilege a single language, Greek, and favor minority students' assimilation of dominant or mainstream values. This chapter argues in favor of a sociopolitical approach to empowerment; this is revisited as locally constructed and interactively accomplished through the orchestration of many different voices. Alternatively, following upon locally emergent perspectives to meaning-making (Canagarajah, 2005), we suggest that research focus be directed toward tracing empowerment as a process shaped in and through the nexus of textual practices constituting a local culture of meanings. Similarly, the processes shaping students' inclusion and exclusion from literate practices are revisited as socially constructed, and are shaped in and through teacher and students' participation in a series of interactive practices created in classroom communities. It is through this classroom culture—consisting of local interactions, emergent identities, and shifting understandings around texts—that students and teachers negotiate and refine their critical stance against centralized meanings.

To clarify the details of this position, we differentiate work on curriculum design and policies concerned with specifying what we henceforth refer to as "school literacy" from work on the co-construction of situated meanings, referred to as "classroom literacies." The former is used in reference to singling out a static, predefined set of linguistic forms and genres proposed as necessary to be acquired by the children. The focus we adopt on classroom literacies suggests that we revisit classroom communities as an emergent habitus containing a multitude of voices competing and even struggling for expression and validation. Indeed, this focus implicates that new meanings be assigned to notions such as "culture" and "classroom culture"—traditionally presented as bounded and static constructs—and new critical questions be raised on the processes involved in their constitution. These questions include: What counts as valued knowledge in this community? What are the processes by which students are shaped as literate subjects? How do local and global processes interact in assigning students the identity of a competent reader or writer?

Drawing upon sociological perspectives to communication, we adopt an approach to culture that moves away from its conceptualization as a "cognitive" phenomenon toward one that stresses its socially constructed nature. Working in the so-called "new sociology of culture," Wuthnow and Witten (1988, 53f) suggest defining culture in terms of discourse and practice. Work in interactional sociolinguistics revisits culture as consisting of "the discourses, texts, symbolic practices, and communicative events that constitute the ongoing stream of social life" (Knoblach, 2001, 25). Restating these proposals in the light of work within Conversation Analysis (Boden and Zimmerman, 1991) that attends to how local realities index and reconstitute social meanings, we suggest that rather than seeing social factors (such as gender and ethnicity), cultural knowledge, and competencies as predefined, fixed, or bounded constructs, should be interpreted as shaped in and through one's participation in the interactions around different kinds of discourses created within communities. According to Gee (1996, 127), discourses are defined as "ways of being in the world, or forms of life which integrate words, acts, values, beliefs, attitudes, and social identities."

While arguing, in line with Gee (1996), that schools forward certain ways of making meanings and reflecting upon texts—captured under the heading of secondary discourses,—a revised notion of secondary discourse competencies is outlined; these are suggested to emerge out of meaning shifts and genre recontextualizations that are proposed to take place within locally emergent curricula units. According to this perspective, empowerment is redefined as an interactional accomplishment constructed out of the intertextual positionings (De Fina et al., 2006) students adopt against both dominant discourses (indexed, among others, by the genres of schooling) as well as against local identities and funds of knowledge.

The question is this: How do students and teachers transform a static reality of linguistic forms and official genres into a multivoiced community that does not privilege certain meaning over others? What appears important for research and pedagogy is to attend to how students are shaped as subjects with certain kinds of competencies rather than others, as well as how competence and expertise are defined across local communities—as the strict reproduction of official meanings or as the exercising of one's agency within global constraints. If the latter is the goal, how can it be accomplished? Relevant research from U.S. contexts has attested that when experiential differences are not treated as deficits (Purcell-Gates, 2002) and children's cultures and experiences are respected and invited in the classroom, children may be willing to add the dominant meanings to their own cultural funds of knowledge (Delpit, 2002). As will be argued below, work on classroom literacies may reveal the processes that shape these acts of meaning invitation and negotiation.

Educational Initiatives for Minority Students: A Brief Overview

The collapse of the former East bloc countries in 1989 resulted in the mass emigration of their citizens. Greece, a country that had exported immigrants to European and transatlantic countries up to the 1970s became an attractive destination for a significant number of economic immigrants, refugees, and repatriates, mainly from East European countries and in particular from Albania and the former USSR. This migration-related diversity had consequences in every sector of public life, including education. Students whose families have come from another country either as immigrants or repatriates attend public schools, visibly changing the demographics in these schools. Currently, the number of students of immigrant or repatriate families in the Greek schools is 10 percent (8.4 percent immigrants and 1.6 percent repatriate children).[1]

In light of this change, in 1996[2] the Ministry of Education initiated the development of programs for all groups of students threatened by exclusion from education or traditionally excluded from it. Starting in 1997, three programs have been developed, all of them cosponsored by the European Union. One of the programs addresses the educational needs of recent arrivals in Greece, while the other two programs focus on indigenous student populations with the highest dropout or nonattendance rates among school age children. The beneficiaries of these programs have been (a) repatriate and immigrant students, (b) Rom children, and (c) Muslim children in Thrace.

Unfolding each one of them, the first program, "Inclusion of repatriate and immigrant children in primary school," started in 1997 and has undergone several phases with periods of nonaction and changes in basic coordinators and contributors. In the school year 2006–2007, it involved 407 teachers and 602 school units, with a school population of 57,550 (both indigenous and nonindigenous students) throughout Greece.[3] Of these, the number of students who were involved in the program in "classes for the inclusion of foreign students," that is, immigrants or repatriates, was 6,988, when for the same school year the total number of immigrant and repatriate students in the country was 130,362 (www.keda.gr). The lack or insufficient dissemination of teaching material is stressed as one of the weaknesses of the program (Papakonstantinou, 2007). Despite the professed intention of enabling the participation of the whole school unit and the promotion of cross-cultural activities, such actions have only been attempted by individual teachers and not as a standard pursuit of all participating school units. The discontinuity of the program and the constant change of the participating teachers and school units is considered as the main weakness of the application of the program, because it "leads to circumstantial and fragmentary implementations and renders the program ineffective, since every time it is reactivated, actions (i.e., teacher training) start almost from the beginning" (Papakonstantinou, 2007, 2).

The second program, the program for the "Inclusion of ROM students in school," also started in 1997 and has also undergone changes in administrative and educational leadership. The majority of Rom is a poor, highly marginalized population, typically living in extreme conditions, at the edges of urban centers. When Rom children go to school they have a very poor command of Greek, since their home language is Romani, a language that has no written form. They are marginalized and sent to usually gravely disadvantaged areas and schools in the country, their school attendance is erratic and their dropout rate is among the highest in the country (Hatzinikolaou and Mitakidou, 2005; Hatzisavvidis, 2007). They are thus greatly underrepresented in secondary education and very few of them make it to tertiary education, thus limiting their chances for social advancement.

The current situation concerning the education of Rom children is rather obscure; data are unreliable and fail to draw the complete picture of the group. A recent attempt to estimate the school attendance of Rom children shows the proportion of children who have never attended school as 50 percent or higher. About 40 percent of the Rom children may have been enrolled in schools, but they attend only the first year of elementary school. Most of those continuing in school also drop out later without acquiring the basic literacy skills in Greek (Hatzisavvidis, 2007).

A common comment for both of the above programs is that they mainly provide for the teaching of Greek as a second language in reception- or support classes. These remedial classes operate with programs designed by experts, as a top-down, given protocol of communication that renders obsolete the children's lived realities, emerging identities, and mother tongue.[4] With the exception of the program for Muslim children, there is no program in any public school in Greece that takes into account the children's native language, let alone endorsing the cultivation of this language. Another common characteristic is that both of these programs run as satellites to the mainstream school, which is the ultimate target, an end in itself for Rom or immigrant and repatriate children. Their native language and home culture are not only insignificant but are often considered obstacles for their schooling.

Interventions informed by the assimilationist ideology are bound to fail because these cannot convince children that they are designed for their best interest when such interventions neglect the basic elements of the children's identity (Delpit, 2002). Expressing his frustration with the long history of ineffective educational programs for Rom children, one of the main contributors of the current program suggests that the school for Rom children should attempt radical changes in its structural, morphological, and functional elements. To be successful, these changes have to be undertaken by the Rom themselves, otherwise they will always resemble the school as it currently is, and this has so far failed to include Rom children (Hatzisavvidis, 2007). Members of the Rom community themselves are aware of this need (Mitakidou and Tressou, 2007).

The third program concerns the education of Muslim students in Thrace.[5] According to the Treaty of Lausanne (1923), Muslim students in Thrace are entitled to bilingual education, and they can attend either mainstream or the so-called minority schools, in which half of the classes are delivered in Greek and half in Turkish. The design and application of this educational scheme has proved ineffective, mainly due to political limitations and barriers. Living in geographical, social, and cultural isolation and being exposed mainly to their home language, Muslim children have seldom completed the nine-year obligatory schooling, since they usually leave primary school with a poor command of linguistic and cognitive skills.

With the motto "addition not subtraction, multiplication not division," reflective of its politics, the program for the education of Muslim children (PEM), has been running without major disruptions in administrative or staff services since 1997. The children participating in the program have been elementary school children (six–twelve years) and their average number in the eleven years of the program's operation has been 7,500 annually (Dragona and Frangoudaki, 2008). The program for the education of Muslim children is part of a general large-scale official effort to reform the political scene in the area and pull the minority population out of isolation. It is an indisciplinary project based on an array of interventions, including teacher training seminars, innovative school approaches (i.e., project, integrated language approaches), development of educational materials, self-assessment, empowerment of families, and so on. Despite the encouraging outcomes,[6] an overall evaluation of its implementation is that the positive impact is greatly annulled by political and ideological forces. This contradiction is due to the very objective of the task, which, as the directors of the program confess, "is the most difficult task in all societies: to disrupt traditional certainties, question self-evident truths, change, in other words, the views and beliefs of people" (Dragona and Frangoudaki, 2008, 52). An added complexity is posed by the fact that changes have been implemented only in the

Greek part of the program, while the Turkish leg has been unaffected, as they remain skeptical and dubious of the attempted reform.

A general impression of this short review of the three programs that have been implemented under the auspices of cross-cultural education is that, despite the ambitious professed goals and the recruitment of many contemporary state-of-the-art practices (though detailed analyses need to be undertaken to reveal the subtleties of these practices), political and ideological barriers have constrained their operation and limited it to support classes for the acquisition of Greek, thus pointing to an ultimately assimilationist practice. As Cummins (2004) points out, the irony is that "language of instruction itself is only the surface structure. Coercive power relations can be expressed as effectively through two languages as through one. Change in the deep structure will come only when educators walk into their classrooms burdened not by the anger of the past and the disdain of the present, but with their own identities focused on transforming the social futures toward which their students are travelling," (p. 10).

To address some of these concerns, we suggest that close attention needs to be paid to the complexities of teacher-students interaction and to the way emerging classroom cultures reconstitute wider social hierarchies.

Literacy Learning in Mainstream Classroom Communities: Redefining Genre Pedagogy for a Diverse Student Population

The pedagogical initiatives on literacy development that have been designed in the Greek context have been following wider research developments on language and literacy. When shifting away from the initially exclusive focus on grammatical and communicative competence to incorporate more recent work stressing genre, generic competence becomes important for defining the nature of school literacies. The curricula and textbooks in use (introduced in 2006 and 2007),[7] through the activities designed, address the genres to be learned by a diverse student population attending elementary school. Textbooks contain series of thematic units within which narrative, descriptive, and argumentative genres are discussed through various topics such as "nutrition," "museums," "holidays," and so on. Although changes in the internal structure of the thematic units may be introduced by teachers (i.e., through the selection of more culturally appropriate texts that respond to the needs of the diverse student population), the result is similarly weak, due to the lack of a more dynamic theoretical framework. Such a framework needs to rest upon redefining genre as a social practice (Hanks, 1996; Kostouli, to appear) and on the ways generic forms may be used to construct and redefine identities and stances toward knowledge.

The clear-cut distinctions of primary versus secondary genres (Gee, 1996) have been currently replaced by proposals arguing for the hybridity of meanings and the embedding and colonization of genres to create larger semiotic structures (Bhatia, 1997). Genres are, thus, no longer seen as static verbal structures transported from context to context but as hybrid multimodal potentialities responding to contextual parameters (Kress et al., 2005; Royce and Bowcher, 2007). These new conceptualizations have significant repercussions on the processes shaping students' routes to literacy learning. Meaning-making is seen not as the reproduction of given meanings but as involving processes of "resourcing the sources" (Stein, 2004, 39). This task cannot be captured by a static set of textbook-inscribed activities. Thus, though retaining Gee's construal of secondary discourses (as indexing specific social languages and identities valued in the school community), we suggest that current inquiry should address more interactive issues: How is the co-construction of the specific school-valued discourse mediated by the discursive events created in classrooms communities? How do students with different learning styles and from diverse backgrounds, through the semiotic resources they draw on, reshape the clear-cut boundaries between primary versus secondary discourses, creating what Lam (2004) refers to as "border discourses"?

Despite the promise of a school that will "no longer be governed by the principles of traditional field-centered curricula that promote a rather passive attitude towards learning...[and] become an institution that promotes joyful and creative living, breaking away from sterile and ineffective teaching practices" (A Cross Thematic Curriculum Framework for Compulsory Education, 6),[8] the new Greek textbooks designed on the basis of this document insist on the pursuit of knowledge through distinct subjects. Though invoking genre as their guiding principle, these textbooks seem to build upon a common premise running through and uniting all programs, mainstream and intervention, alike: empowerment is seen as a decontextualized construct that appears in a similar manner across student populations and communities, that is, through the appropriation of the so-called genres of schooling or the genres of power (Schleppegrell, 2004).

Drawing upon notions, such as recontextualization and meaning shifts across genres, and integrating premises drawn from interrelated fields within a Bakhtinian-inspired framework on dialogic language use, this chapter attends to the discursive construction of school genres and classroom literacies. Rather than taking secondary discourses as fixed powerful resources, these are seen as a sequentially emergent web of meanings, identities, representational resources, and stances toward knowledge constructed through the orchestration of different modes; indeed, this orchestration is proposed to reflect and indicate specific cultural ideologies and societal meanings (Kamberelis, 2001).

Within this perspective, written texts are seen as only a part of a complex network of semiotic resources that contribute to the construction of a deeply political orientation to language use and "link us to particular social histories and their supporting institutions" (Duranti, 1997, 49). In line with Critical Discourse Analysis, we suggest that analysis of writing resources is important for unveiling the ideologies shaping classroom activities and the ways students are constructed and repositioned as specific kinds of subjects.

Conceptualizing Empowerment as Textually and Interactionally Shaped Basic Premises

The way language relates to and constructs social realities and practices that empower or disempower students constitutes the basic issue attended to by social constructionism—an umbrella term used for a range of theories (drawing on Marxism, the Frankfurt school, Habermas, Bakhtin, critical theory) that explore the relation between culture, reality, social life, knowledge, and meaning-making. Specific strands have developed in fields such as Critical Discourse Analysis (CDA), Mediated Discourse Analysis (MDA), Conversation Analysis (CA), and discursive psychology, which may differ in a number of respects. Despite such differences, these research strands basically subscribe to an antiessentialist position to language, knowledge, and indeed to the social world. Power relations, ideologies, cultural models, and social identification categories (race, gender, etc.) are presented to be "interactionally constructed, socially transmitted, historically sedimented and...communicatively reproduced in situ" (Gunnarsson, Linell, and Nordberg, 1997, 2).

Alternatively, social practices are understood as being constituted in and through discursive social interaction, while, at the same time, locally emergent social interactions are seen as suggesting renegotiations of social practices—that is, of the meanings and the positionings to be taken up by participants. Thus, in contrast to social theory attending to the way habitus shapes subject positions, people are not presented to just assume the subject positions interpellated to them by discourses; they are "acting within meanings located at the level of socio-history...and within meanings situated at the level of local social interaction" (Moita-Lopes, 2006, 293).

What are the implications of this approach to classroom-situated multicultural research on literacy? Let us begin by noting that while the role schools play in reproducing wider social structures has been a prevailing theme in sociological research, and in the last few years attempts have been made to connect sociolinguistics with social theory (see Erickson, 2004), genre-based literacy research has not engaged in such a dialogue. The way in which the ecologies of textual practices constructed in classrooms relate to, reflect, index or restructure wider resources and discourses remains a poorly researched topic. In pursuing this topic, important issues arise.

How can we move from tracing the linkages between single texts and ideological processes (as is typical in CDA) toward illustrating the way a series of constantly evolving contexts establishes a locally-valued perspective? How does this perspective signal the role classroom communities play for reproducing versus rearticulating social practices? Building upon Bakhtin (1981, 1986), we suggest that a communicative landscape is created by participants interweaving "language drawn from and invoking alternative, cultural, social and linguistic home environments" and by the teacher and other peers managing "the interpenetration of multiple voices and forms of utterance" (Duranti and Goodwin, 1992, 19).

How is writing and genre production to be reconceptualized as a tool empowering students in classroom communities? To capture this, analysis should shift away from proposals of writing development as an individual accomplishment. Emphasis should be placed on the way in which multiple meaning trajectories are dialogically co-constructed through whole-class and peer-group settings (Blommaert, 2005).

EXTENDING CRITICAL DISCOURSE ANALYSIS TO CLASSROOM-SITUATED GENRE RESEARCH

To understand the political nature of classroom-situated literacy processes, we proceed beyond CDA (Fairclough, 1989, 2003, 2004; van Dijk, 1998; Weiss and Wodak, 2003) and develop a more elaborate account that captures the way students' texts and classroom processes mediate the emergence of school-valued genre competencies. Building upon Mediated Discourse Analysis (MDA), a research field that extends CDA through Wertsch's (1998) "sociocultural theory of mind," we suggest that the focus should no longer be on single texts or discourses as such but rather on mediated action. Language is an important mediational tool people use along with other tools (each embodying certain affordances and constraints) to take action in a polyphonic and heteroglossic community. MDA attempts to explain how various mediational means, including discourses, reproduce and transform discourses. MDA sees discourse as "cycling" through social actions (Norris and Jones, 2005; Fairclough, 2006). MDA shares with CDA a concern with hybridity and an interest "in the ways multiple 'voices' interact in interdiscursive dialogicality" (Jones and Norris, 2005, 9), though the role of discourse is downplayed. Mediated action is taken as the unit of analysis.

Cultural Context/Cultural Models and Ideologies

Figure 3.1 (a revised version of Kostouli, 2005) and Figure 3.2 present two gradually refined ways of conceptualizing the way the premises of CDA and MDA may inform aspects of language use and learning in classroom communities.

Extending Fairclough's work, in Figure 3.1, texts and interactions are seen as units of social action that acquire meaning by being embedded in a social context and reflecting a historical order of meanings (Blommaert, 2005).

The notion of "genre sets" (Bazerman, 1994) in Figure 3.2 helps us capture the dynamic nature of classroom-situated meanings, the ways in which local histories of meanings emerge and get reshaped in

Figure 3.1   Situating texts against contextual layers

Figure 3.2   Genre sets against contextual layers

classroom communities. The notion of genre sets is used after Smart (2006, 202) in reference to the full repertoire of genres—presented in the form of written texts and oral processes, constructed by the teacher and the students as they collaborate to perform specific tasks and attain certain goals. The construction of genre sets is part of an ideological work any given community engages in, which is part of the communicative processes by which the social stocks of knowledge are being built up, maintained, transmitted, and modified (Gunthner and Knoblauch, 1995, 5). Given Figure 3.2, a number of issues emerge for cross-cultural research:

- How do teacher and students, through the texts students produce and the units they co-construct within and across genre sets, bring into the classroom conversational floor competing voices?
- How do they integrate these into an interaction that leads to a "shared" culture of meanings?
- How do teachers and students negotiate and restructure competing discourses on gender, race, and social class?
- How are the identities of students as specific kinds of learners shaped through this culture of meanings?
- How do children appropriate or resist these identities through the texts they produce and/or the way they participate in this local culture of meanings?
- How can the results of this work help teachers revisit their practices that constrain students' identities and academic achievement?

LEARNING TO WRITE IN SCHOOL GENRES

The notion of genre sets has been important in our own work on genre learning in Greek elementary classrooms (see Kostouli, 2005b, 2006, to appear, for mainstream classrooms with a diverse student population; Mitakidou and Tressou, 2005 for reception classes). Kostouli focuses on the way teachers cooperated throughout the year with their students (from 4th grade onward) to implement the centralized curriculum versus a new curriculum characterized by a dialogic and critical orientation to knowledge. By creating and implementing the program "Teaching language and mathematics through literature" in reception classes, Mitakidou and Tressou explore the dynamics of acquiring a second language through its use for authentic communication within social contexts.

While an extensive review of this work is not possible in this chapter, a brief outline of the findings is necessary for indicating the way the theoretical notions outlined in previous sections may be translated in actual classroom practices. Rather than focusing on the development of speaking, writing, listening, and reading as autonomous skills, attention is directed to the meaning-making practices constructed (through the genre sets developed) within classroom communities. Indeed, classrooms were found to be differentiated into those concerned with "bringing the outside in" versus those operating as contexts of authoritative meanings.

CONCLUSION

In this chapter, it has been proposed that learning in multicultural classroom contexts cannot be analyzed independently of the processes through which wider social hierarchies, power relations, and subject positions are taken up and negotiated through various mediational means in classroom communities. Rather than seen as a harmonious process, learning may involve struggles over access to representational resources and competing viewpoints. What about teaching?

Further research across diverse contexts may offer useful insights of the nature of the dialogic processes teachers across classroom communities employ to reshape local meanings toward a more internally persuasive manner. Research on revising classroom practices cannot be limited to the use of a specific textbook or local strategies assumed to have a dialogic function. Teacher education should bring to the foreground the complexity of teachers' interactions with students and the significant role local discourses play in shaping new classroom positionings against global societal hierarchies. Further research is needed on how the insights of these theoretical orientations may be translated in teacher education courses that develop teacher reflexivity regarding their role in co-constructing communities of learning that redefine wider social asymmetries and support pedagogies of inclusion.

Notes

1. The statistics are from the 2006–2007 school year. Source: IPODE (Institute for the Education of Repatriate Children and Cross-Cultural Education). http://www.ypepth.gr/el_ec_home.htm
2. Law 2413/96 defines policies and educational schemes in the framework of "Cross-cultural education" (the Ministry's term).
3. These are schools mostly in the vicinity of major Universities in six urban centers.
4. For the operation and a critique of these classes, see Mitakidou, Tressou, and Daniilidou (2007).
5. A mixed student population consisting of a majority of Turkish speaking children, a smaller group with students speaking Pomak, a Slavic language, and an even smaller group of Rom students speaking Romani. Despite different home languages, however, all groups have some proficiency in Turkish.
6. In the eleven years of its implementation, the number of Muslim children continuing onto secondary education has quadrupled. Attendance in the Lyceum (senior high school, not obligatory) has risen by 60 percent in the past four years and the dropout rate has decreased from 65 percent in 2000 to its half in 2007. Of course the dropout rate is much higher than the national average (7 percent) but the improvement is significant (Dragona and Frangoudaki, 2008).
7. The Greek educational system is a totally centralized one, so curricula and textbooks are distributed and implemented in every school in the country.
8. Diathematiko Enieo Plaisio Programmaton Spoudon. [Cross Thematic Curriculum Framework for Compulsory Education]. Translated from the official gazette issue b, nr 303/13-03-03 and issue b, nr 304/13-03-03 by members of the P. I. (Pedagogical Institute) main staff and teachers seconded to the P. I. http://www.pi-schools.gr/programs/depps/index_eng.php (accessed July 19, 2008).

References

Bakhtin, M. M. (1981). *The Dialogic Imagination*. (Trans. C. Emerson and M. Holquist). Austin: University of Texas Press.

——. (1986) *Speech Genres and Other Late Essays*. (Trans. V. W. McGee; ed. C. Emerson and M. Holquist). Austin: University of Austin Press.

Bazerman, C. (1994). "Systems of Genre and the Enactment of Social Intentions." In A. Freedman and P. Medway (eds.), *Rethinking Genre*. Madison: University of Wisconsin Press. 79–101.

Bhatia, V. K. (1997). "Genre-mixing in Academic Introductions." *English for Specific Purposes*, 16, 181–195.

Blommaert, Jan (2005). *Discourse: A Critical Introduction*. Cambridge: Cambridge University Press.

Boden, D. and D. H. Zimmerman (eds.) (1991). *Talk and Social Structure: Studies in Ethnomethodology and Conversation Analysis*. Cambridge: Polity Press.

Canagarajah, A. S. (2005). *Reclaiming the Local in Language Policy and Practice*. Mahwah, NJ: Lawrence Erlbaum Associates.

Cummins, J. (2004). *Language, Power and Pedagogy: Bilingual Children in the Crossfire*. Clevedon: Multilingual Matters.

De Fina, A., Schifrin, D., and Bamberg, M. (eds.) (2006). *Discourse and Identity*. Cambridge: Cambridge University Press.

Delpit, L. (2002). "No Kinda Sense." In L. Delpit and J. K. Dowdy (eds.), *The Skin That We Speak*. New York: The New Press. 31–48.

Dragona, Th. and Frangoudaki, A. (2008). *Prosthesi ohi Aferesi. Pollaplasiasmos ohi Dieresi.* [Addition not subtraction. Multiplication not division]. *Athens: Metehmio.*

Duranti, A. (1997). *Linguistic Anthropology*. Cambridge: Cambridge University Press.

Duranti, A. and Goodwin, C. (eds.) (1992). *Rethinking Context: Language as an Interactive Phenomenon*. Cambridge: Cambridge University Press.

Erickson, F. (2004). *Talk and Social Theory: Ecologies of Speaking and Listening in Everyday Life*. Cambridge: Polity Press.

Fairclough, N. (1989). *Language and Power*. London: Longman.

———. (2003). *Analyzing Discourse: Textual Analysis for Social Research*. London: Routledge.

———. (2004). "Semiotic Aspects of Social Transformation and Learning." In R. Rogers (ed.), *An Introduction to Critical Discourse Analysis in Education*. Mahwah, NJ: Lawrence Erlbaum Associates. 225–235.

Gee, J. P. (1996). *Social Linguistics and Literacies: Ideology in Discourses*. 2nd edition. New York: Falmer Press.

Gunnarsson, B.L., Linell, P., and Nordberg, B. (eds.) (1997). *Professional Discourse*. London: Addison Wesley Longman.

Gunthner, S. and Knoblauch, H. (1995). "Culturally Patterned Speaking Practices-The Analysis of Communicative Genres," *Pragmatics*, 5(1), 1–32.

Hanks, W. F. (1996). *Language and Communicative Practices*. Boulder, CO: Westview Press.

Hatzinikolaou, A. and Mitakidou, S. (2005). "Roma Children: Building Bridges." In L. D. Soto and B. B. Swadener (eds.), *Power and Voice in Research with Children*. New York: Peter Lang.

Hatzisavvidis, S. (2007). *Eterotita sti scholiki taksi ke I didaskalia tis ellinikis glossas ke ton mathimatikon. I periptosi ton tsigganopedon* [*Diversity in the Classroom and the Teaching of Greek and Mathematics. The Case of Rom Children*]. Volos: YPEPTH and Panepistimio Thessalias.

Jones, R. H. and Norris, S. (2005). "Discourse as Action/Discourse in Action." In S. Norris and R. H. Jones (eds.), *Discourse in Action: Introducing Mediated Discourse Analysis.* London: Routledge. 3–14.

Kamberelis, G. (2001). "Producing Heteroglossic Classroom (Micro)cultures through Hybrid Discourse Practice," *Linguistics and Education,* 12, 85–125.

Knoblach, H. (2001). "Communication, Contexts and Culture: A Communicative Constructivist Approach to Intercultural Communication." In A. Di Luzio, S. Günther, and F. Orletti (eds.), *Culture in Communication: Analyses of Intercultural Situations.* Amsterdam: John Benjamins. 3–33.

Kostouli, T. (2005). "Introduction: Making Social Meanings in Contexts." In T. Kostouli (ed.), *Writing in Context(s): Textual Practices and Learning Processes in Sociocultural Settings.* Boston: Springer. 1–26.

———. (to appear). "A Sociocultural Framework: Writing as Social Practice." In R. Beard, D. Myhill, and J. Riley (eds.), *The Handbook of Writing Development.*

Kress, G., Jewitt, C., Bourne, J., Franks, A., Hardcastle, J., Jones, K., and Reid, E. (2005). *English in Urban Classrooms: A Multimodal Perspective on Teaching and Learning.* London: Routledge Falmer.

Lam, W. S. E. (2004). "Border Discourses and Identities in Transnational Youth Culture." In J. Mahiri (ed.), *What They Don't Learn at School: Literacy in the Lives of Urban Youth.* New York: Peter Lang. 79–97.

Mitakidou, S. and Tressou, E. (2007). *"Na sou po ego pos tha mathoun grammata" Tsiganes miloun gia tin ekpedefsi ton pedion tous.* [*"Let Me Tell You How They'll Learn to Read and Write." Gypsy Women Talk about the Education of Their Children*]. Athens: Kalidoskopio.

Mitakidou, S. Tressou, E., and Daniilidou, E. (2007). "Cross-cultural Education: A Challenge or A Problem?" In M. Bloch and B. B. Swadener (special guest editors), *Education for All: Social Inclusions and Exclusions. International Journal of Educational Policy, Research & Practice: Reconceptualizing Childhood Studies.* 67–81.

Moita-Lopes, L. P. (2006). "On Being White, Heterosexual and Male in A Brazilian School: Multiple Positionings in Oral Narratives." In A. De Fina, D. Schifrin, and M. Bamberg (eds.), *Discourse and identity.* Cambridge: Cambridge University Press. 288–231.

Norris, S. and Jones, R.H. (eds.) (2005). *Discourse in Action: Introducing Mediated Discourse Analysis.* London: Routledge.

Papakonstantinou, G. (2007). *Entaksi pedion palinnostounton ke allodapon sto sholio—Gia tin protovathmia ekpedefsi.* [*Inclusion of Repatriate and Foreign Children in School—for Primary Education*]. Athens: YPEPTH and Ethniko Kapodistriako Panepistimio of Athens.

Purcell-Gates, V. "'…As Soon as She Opened Her Mouth!': Issues of Language, Literacy, and Power." In L. Delpit and J. K. Dowdy (eds.), *The Skin That We Speak.* New York: The New Press. 121–141.

Royce, T. D. and Bowcher, W. (eds.) (2007). *New Directions in the Analysis of Multimodal Discourse.* Mahwah, NJ: Lawrence Erlbaum Associates.

Schleppegrell, M. J. (2004). *The Language of Schooling: A Functional Linguistics Perspective.* Mahwah, NJ: Lawrence Erlbaum Associates.

Smart, Gr. (2006). *Writing the Economy: Activity, Genre, and Technology in the World of Banking.* London: Equinox.

Stein, P. (2004). "Re-sourcing Resources: Pedagogy, History and Loss in a Johannesburg Classroom." In M. R. Hawkins (ed.), *Language Learning and Teacher Education: A Sociocultural Approach.* Clevedon: Multilingual Matters. 35–51.

Street, B. V. (ed.) (2005). *Literacies across Educational Contexts: Mediating Learning and Teaching.* Philadelphia: Caslon.

Van Dijk, T. A. (1998). *Ideology: A Multidisciplinary Approach.* London: Sage.

Weiss, G. and Wodak, R. (2003). *Critical Discourse Analysis: Theory and Interdisciplinarity.* Basignstoke: Palgrave Macmillan.

Wertsch, J. (1998). *Mind as Action.* New York and Oxford: Oxford University Press.

Wuthnow, R. and M. Winen (1988). "New Directions in the Study of Culture." *Annual Review of Sociology,* 14, 49–67.

CHAPTER 4

Pedagogies of Inclusion: Lessons from *Escola Cidadã*

Gustavo E. Fischman and Luis Armando Gandin

INTRODUCTION

Located in the south of Brazil, Porto Alegre is the largest urban district and the capital of the State of Rio Grande do Sul. The city has a population of 1,400,000 and from 1989 to 2005 it implemented the *Escola Cidadã* (Citizen School Project), an educational reform project.[1] *Escola Cidadã* stands out as one of the most innovative urban educational reform projects implemented worldwide (UN, 1996; World Bank, 2000) because it has been successfully educating children and the citizenry of Porto Alegre, and doing so under severe economic constraints and against the rationale of the dominant models of educational reform (Fischman, Ball, and Gvirtz, 2003). In this chapter we will argue that *Escola Cidadã* has not only improved the quality of education in the city of Porto Alegre, Brazil, but it has also the potential for becoming a worldwide reference for how we could think about the politics of educational policy and social exclusion.

This chapter is organized as follows: section one presents a brief discussion of educational reform in Brazil during the 1980s and 1990s as a way of contextualizing the radical departure of the Citizen School Project and its emphasis on participation and democratization. Section two describes the three main goals and structures of the Citizen School Project: democratization of access to schools; democratization of schools' administration; democratization of access to knowledge. Section three concludes this article by discussing the key dynamics through which the educators and citizens of Porto Alegre are engaging in the complex process of building a school that is academically, socially, and politically significant.

The Context of the Citizen School Project: Educational Reforms in Latin America

The vast literature that addresses the contemporary changes in educational systems associated with the processes of globalization and capitalistic reorganization demonstrates above all that formal systems of schooling are positioned as one of the most important social policies of the state (Anyon, 2005; Ozga, 2000; Fischman et al., 2005).

Echoing reforms in the United States, the UK, Canada, and Australia, many of the efforts aimed at reforming education in Latin America during the 1990s emphasized discourses in which the concepts of choice, accountability, efficiency, and decentralization figured prominently, signaling to societies and educational communities the notion that schools should operate with more autonomy, following a "business model" (Hursh, 2006). Many of these reforms had the paradoxical effect of reducing an already small degree of school autonomy, enabling central governments to acquire even greater control over the daily life of educational institutions. In Latin America, after several reforms aimed at increasing school autonomy, federal and state departments of education have even more power to determine school policies, curriculum, and evaluation. Moreover, highly centralized agencies have been created with the means to exercise tighter control over the performance of individual schools, punishing and rewarding them (Klees, 2007).

The most alarming dynamic of this new type of control is that academic performances are not driven by educational principles or needs, but are established to meet the goals of financial programs of economic adjustment (better known as Structural Adjustment Programs, SAPs or, more recently, Poverty Reduction Strategies). The implementation of SAPs, devised and supervised by international financial institutions, such as the World Bank and the International Monetary Fund, has required a significant reduction in the size of the state and public sector as well as in its regulatory functions. This retrenchment of the state has been extended to many social services, including education, health, and related forms of welfare; and in most cases SAPs have exacerbated the gap between the rich and the poor and created important challenges in the provision of basic services that were previously guaranteed by the state (Reimers, 2006, 2007).

In the particular case of Brazil, the educational policies of the 1990s also followed the conceptual guidelines and constraints related to the implementation of SAPs. It is important to acknowledge that during this period the federal and state governments expanded the number of buildings, teachers, and students; yet, what remains a source of great controversy is the actual relevance of those educational policies in terms of equity and for improving the educational outcomes of poor children. The improvement in the enrollment numbers per se guarantees neither better opportunities in terms of jobs (because there is no guarantee that the students that entered schools will continue with schooling) nor an education that can contribute to changing the immense inequality in Brazil. When you add more students to a system that maintains its tendency of disrespecting the culture of the student and see him or her simply as another individual who can be "trained" in the basic abilities necessary to the world of paid work, you create another problem rather than solving the original one. Who will guarantee that these students will not drop out of school, as they historically have been doing in Brazil?

Another important aspect that should be stressed is that education in Brazil, as a rule, is centralized. In the majority of states and cities there are no elections for the city or state council of education (traditionally a bureaucratic structure, with members appointed by the executive), let alone for principals in schools. The curriculum is usually defined by the departments of education of cities and states. Since the resources are administered by centralized state agencies, schools usually have very little or no autonomy.

Although recently Brazil has achieved a very high level of initial access to schools (close to 98 percent), the indexes of failures and dropouts are frightening. Brazilian data shows that historically less than 70 percent of the students in first grade passed to second grade in their first attempt. Hence, although initial access is granted, the chance of a poor child passing to second grade is low. Furthermore, the dropout rate is extremely high, close to 20 percent, in fourth grade. These data show clearly the exclusion being reproduced in the Brazilian educational system.

This brief description of the main features of educational reform in Latin America and the general performance of schooling in Brazil provides a context for understanding the enormous educational challenges the citizens of Porto Alegre had to face while implementing the Citizen School Project.

REEDUCATING THE CITY OF PORTO ALEGRE

From its inception in 1989, the Citizen School Project has been considered an educational utopia because it is organized around three main principles: (1) schooling is relevant and meaningful if it provides individual and social opportunities for learning to read the word and the world (Freire, 1982); (2) learning to read the word and the world is only possible when the educational community in general and educators in particular recognize the interdependence of the political dimensions of urban schooling and the pedagogical dimensions of political participation in the urban landscape; and (3) democratizing schools requires individual and collective efforts to create an educational project that is open and flexible in its structures while maintaining its goals of radically democratizing school practices.

Within the Citizen School Project, schools are transformed into laboratories for the practice of individual rights and social rights. At the center of this project are the following ideals: encouraging the development of autonomous, critical, and creative individuals; nurturing a model of citizenship that supports daily practices of solidarity, justice, freedom, and respect for diversity; informing all curriculum practices with a commitment to the development of a less exploitative relationship with the environment. José Clóvis de Azevedo, former Secretary of Education of Porto Alegre, states that to achieve such ambitious goals it is necessary to recover the sense of schools as laboratories of democracy, a notion that directly counters technocratic attempts to run schools following commercial or market oriented models:

> We reaffirm our commitment to expand the humanist character of public schools and we oppose the submission of education to the values of the market and neo-liberal reforms in education. The market's main concern is to form consumers and customers, to turn education into merchandise submitted to profit-seeking rationales, naturalizing individualism, conformity, unfair competition, indifference and, consequently, the exclusion of those deemed a-priori unsuited to compete. (Azevedo, 1999a: 5. Our translation)

The educational reforms created in Porto Alegre and envisioned in Azevedo's statement are still a work in progress, and is based on a set of core principles. One of those principles is that truly inclusive schooling will not result from simply changing technical, administrative, and pedagogical aspects. Rather, new teaching methods, organizational changes, administration structures, and leadership must be coupled with a global political project as an effort to enact more democratic power relationships. Thus, the Citizen Schools Project is an ambitious project of educational transformation, which combines pedagogical innovation with a shift in the type of relationships cultivated within the school setting and the type of knowledge being taught.

To implement this transformation, the coordinators in the Municipal Secretariat of Education (SMED) knew that they would have to incorporate many students (mostly the children of migrant families living in the poorest areas of the city) who had been excluded from the educational system; in short, they would have to build new schools. The number of schools increased from 29 in 1988 to 92 in 2004. In 1988, the municipal school system of Porto Alegre provided educational services to only 17,000 students[2] and had a dropout rate of almost 10 percent. In 2004 the system increased the coverage to 60,000 students, and the dropout rate fell to 2 percent. The number of teachers also increased dramatically: in 1988 there were 1,700 teachers, and in 2004 the number of teachers in municipal schools increased to 4,000. These teachers were among the best paid in public educational systems in Brazil, earning three times what a state teacher earned. Other indicators of the quality of this reform project are changes in curriculum and organizational structures that address the needs of a diverse urban population; high levels of community participation in the life of the schools; and the key role that schools play in the lives of local communities.

These are impressive results for a Brazilian school district, one that had been surrounded and pressured by other educational models that operated from a completely different perspective. Furthermore, the results have been achieved following the core values of empowerment, collective work, respect for differences, solidarity, knowledge as a historical experience, and citizenship aimed at democratizing access to school, knowledge, and governance. Given these core values, since its conception and initial implementation in 1989, the Citizen School Project has been explicitly designed to radically change both the municipal schools and the relationship between communities, the state, and the educational system. This set of policies and the efforts to implement them are constitutive parts of a clear and politically explicit project aimed at developing institutions that will provide opportunities for not only better schooling for the students who have been traditionally excluded, but also the provision of knowledge, opportunities for practice, and structures expected to support the constant expansion and strengthening of a social project of participatory democracy.

The Citizen School Project is not an isolated initiative. It is part of a larger program of radical transformation of the relationship between the state (in this case, the municipal state) and civil society. Perhaps the best known of those initiatives is a system of participatory decision-making program used to allocate up to one-third of the funds from the municipal budget (*Orçamento Participativo* or "OP"). The OP is a system that allows citizens to decide about the city's budget provision in the municipality. Regular issues discussed in the OP model included improvements at the level of a single neighborhood or district, such as pavements, sewages, storm drains, schools, health care, childcare, housing, and so on.[3]

Several studies (Abers, 2000; Baierle, 1998; Genro, 2001; Santos, 1998; Schugurensky, 2001) have concluded that the OP has not only encouraged more transparency and efficiency in the allocation of public resources but has also improved the living conditions of poor communities by redirecting tax revenues to previously neglected areas. As the prominent Portuguese political scientist Boaventura de Sousa Santos noted:

> The participatory budget promoted by the Prefeitura (government) of Porto Alegre is a form of public government that tries to break away from the authoritarian and patrimonialist tradition of public policies, resorting to the direct participation of the population in the different phases of budget preparation and implementation, with special concern for the definition of priorities for the distribution of investment resources. (Santos 1998, p. 467)

The OP is at the core of the project of transforming the city of Porto Alegre and incorporating a historically excluded impoverished population into the processes of decision-making. As we noted before, not only have the material conditions of the impoverished population improved, but also the OP has generated an educative process that has forged new social organizations and associations in neighborhoods, and it is intimately related to and complements the Citizen School Project decision-making process, because the OP constitutes one of the organizing structural bases for hope.[4]

The Citizen School Project

One of the most salient characteristics of the *Citizen School Project* is that it is organized around participatory structures that encourage educators, students, parents, community organizations, and individuals to participate in the decision-making process about the role that schools should play in the larger society. These decision-making processes require important levels of reflection upon the type of social, political, and educational practices they would like to see in operation in the municipality's schools. It is worth highlighting that in 1989, even amid a dramatic national financial and economic crisis, the Citizen School Project expressed the articulation of democratic ideals, community experiences, the legacy of the popular education movement, and a firm commitment to create a new model of schooling. The educational policies implemented subsequently by four different administrations of the Secretary of Education of the City of Porto Alegre infused a radical democratic spirit into the educational sphere by supporting the direct and active participation of students, teachers, administrators, staff, parents, and the community at large in the formulation, administration, and control of the municipal public policies.

In 1993, the Citizen School Project implemented a series of regional meetings designed to prepare a collective structure open to the participation of all the segments of the school community, which was called "School Constituent Congress." The goal of the School Constituent Congress was to establish the guiding principles that would orient the construction of an inclusive, democratic, and emancipatory school. The School Constituent Congress was attended by elected delegates, guaranteeing the participation of parents, students, teachers, and staff. The Congress proclaimed that the democratization of schooling was the main goal of the Citizen School Project, and in order to achieve such goal the project had to create the structures that would guarantee democratization of access to schools, democratization of schools' administration, and democratization of access to knowledge. The next sections will expand and provide examples of what democratization of schools means for the Citizen School Project.

Establishing the Bases for Inclusion: Democratizing Access to Schools

As noted before, for the organizers of the *Citizen School Project*, the democratization of access to schools was a priority and explicitly connected to the educational needs of the most impoverished neighborhoods of Porto Alegre where the municipal schools are situated. For the Popular Administration, guaranteeing access to public schooling for children of communities historically neglected by the state and its network of social services was the first step toward promoting social justice.

Accordingly, one of the first curricular and organizational reforms instituted to address the social and educational exclusion suffered by many students in Porto Alegre was abolishing the grade structure and establishing a "cycle of formation" system. The three cycles that were thus established are each three years long, totaling nine years of education.[5] The education by cycles attempts to eliminate school-related mechanisms that perpetuate the exclusion of mostly poor and racialized minority students through failure and dropping out. The student's progress from one year to the other and the notion of "individual educational failure" was eliminated. The latter is meant to reframe the conception of failure present in many educational systems that traditionally "solve educational failure" by blaming the student (or her/his family, or his/her race, culture, linguistic ability, social class, etc.) for her/his school performance, increasing the likelihood of the student dropping out of schools in later grades, without examining the role of the school in creating the very notion of student failure (Shepard and Smith, 1989). The fact that the concept of student's learning failure was eliminated does not mean that assessment of the learning process of the students is not done. This assessment is very rigorous and teachers organize dossiers that accompany students from one year to the next in order to help the next teacher to emphasize aspects that were not optimal in the learning process of the student.

The implementation of the cycles (Krug, 2001) as part of the process of elimination of school-related mechanisms of social and educational exclusion, though important, was not enough. The Citizen School also created several mechanisms that guaranteed the re-inclusion of students who had previously been excluded from the school system. One of those is the implementation of progression groups and learning laboratories. In progression groups, students who come into a Citizen School from other school systems (e.g., state schools) after having experienced multiple failures are provided special attention until they are integrated into a cycle at the appropriate educational level. Learning laboratories are not only a space where students with special needs are helped, but also a place where teachers conduct research activities (in the teacher as researcher model, Anderson, Herr, and Nihlen, 1994; Zeichner and Noffke, 1998; Gandin, 2004) in order to improve the quality of the regular classes.[6] A major emphasis of these mechanisms was the transformation of the school structure to improve inclusion, participation, and learning.

Finally, the material conditions of the schools are excellent. All the schools are kept in prime condition, with good libraries, lighting, cooking facilities, and playgrounds. Almost all schools also have computer labs, gymnasiums, dance floors, or painting and sculpture studies. For those in the United States or Europe, most of these facilities are a given in public schools, but in Brazil their availability is remarkable and is a key component in strengthening the bases for hope by guaranteeing access and permanence of students at the municipal schools of the city of Porto Alegre.

Strengthening the Bases for Inclusion: Democratizing the Administration of Schools

The School Constituent Congress mandated the establishment of mechanisms for the direct election of principals and assistant principals in the schools of Porto Alegre. This mechanism aimed at the redefinition of power relationships inside schools. To be elected to occupy positions of leadership, candidates are required to present a feasible pedagogical and administrative program for the school. The proposed program has to state pedagogical goals and the means to achieve them; must obtain the support of staff, teachers, students, and parents; and has to be technically and financially responsible. In this manner, the goals, procedures, and norms for administrative and pedagogic relationships that are developed specifically for a given school require a process of consensus building through informed dialogue among teachers, students, educational authorities, parents, and members of the local community.

Another vital mechanism developed to democratize the exercise of power inside schools was the creation of School Councils, which comprised elected representatives (parents, students, employees, and teachers). The School Council is the highest decision-making level of the school, and it exercises considerable influence over administrative, financial, and pedagogic matters. The Council's main responsibility is to define the general orientation of the school, to determine the administration's guidelines, and to allocate the school's resources. The school principal, who is a member of the Council, is also responsible for offering political and pedagogical coordination to the general project (Azevedo, 1999b).

These mechanisms of governance have impacted not only the administrative aspects of school life but also the sense of professionalism among teachers and principals, and perhaps, in a crucial manner, have opened real opportunities for deepening the bases for hope by emphasizing the importance of democratizing the access to knowledge.

Deepening the Bases for Inclusion: Democratizing the Access to Knowledge

Curriculum transformation is another crucial goal of Porto Alegre's project for transforming schools through more democratic practices. Curricular transformation is more than making sure that the students are going to be offered access to traditional knowledge required for an educated and enlightened citizenry. The Citizen School Project goes beyond the incorporation of new knowledge within the margins of an intact "core of humankind's wisdom." It is a radical transformation aimed at constructing a new epistemological understanding about what counts as knowledge. Thus, in these schools, teaching goes beyond the mere episodic mentioning of the structural and cultural manifestations of class-, racial-, sexual-, and gender-based oppression. It includes these themes as an essential part of the process of construction of knowledge.

Perhaps no other element better illustrates the meaning of what teaching is for the Citizen School Project than the role of the educator as defined by the School Constituent Congress:

> The educator's role is to be next to the student, challenging the real and imaginary worlds brought to school by students, contributing to the life-world of the students in such a way that the "world" can be understood and reinvented by the student. The educator should also grow, learn, and experience together with the students, the conflicts, inventions, curiosity, and desires, respecting each student as a being that thinks differently, respecting each student's individuality. (SMED, 1999, 57. Our translation from Portuguese)

This program helps consolidate the mutual responsibility and obligations of the state and civil society, teachers, communities, and students by reinventing the concept of "schools for citizenship" where "achievement for all" is only the first step toward creating more democratic spaces in education. As John P. Myers (2007) has concluded in his study about teachers in Porto Alegre:

> Rather than preparing students for minimal civic participation or abstract knowledge of democracy, their teaching addressed the context-specific challenges to improving democracy in their communities

and nations. The teachers in Porto Alegre collectively emphasized the problem of unequal social-class relations, which was influenced by several sources: their involvement in class-based parties, the tradition of popular education, and (for some) their students' life experiences.

Tarso Genro (a former mayor of Porto Alegre and an ex-Minister of Education under the Lula government) states:

> The truth *about* Escola Cidadã is that fundamentally it is a rational and undetermined space. In it, citizens—teachers and learners—are connecting with the values and legacies of enlightenment, tolerance, respect for human diversity and cultural pluralism, in dialogical coexistence, sharing experiences and knowledge and identifying history as open future. (Genro 1999: 11. Our translation)

Curriculum in the schools of Porto Alegre is constructed in a decentralized manner, with participatory research strategies being the starting point. One of the most effective tools for resolving the frequent disconnection between the cultural and social frameworks of communities and schools is the use of educational thematic units built around a central concern for the community. Teachers develop these units after conducting a social-anthropological diagnostic evaluation within their school communities.

These thematic units become locally based and locally owned instruments designed to construct and distribute knowledge that is socially relevant for the communities served by each school. The critical incorporation of elements deemed relevant for the school community in addition to pedagogical practices developed to strengthen the concept of radical democracy within *Escola Cidadã* have given a new meaning to teaching and learning in Porto Alegre.

The starting point for this new process of knowledge construction is the idea of "thematic complexes." Through research (one that the teachers do in the communities where they work, involving students, parents, and the whole community), the main themes from the interests or worries of the community are listed. Then the most significant one is defined as the focal point of the thematic complex. From this focal point, each knowledge area defines the crucial concepts (and from these, content) that can help study the issue in question. Therefore, the curriculum is organized around the focal point, in a thematic complex that will guide the action of the classroom, in an interdisciplinary form, and during a period of time.

For example, one of the schools defined its focal point as "quality of life." All the teachers of different knowledge areas work around this focal point to offer their specific concepts as a contribution to a complex understanding of this issue. In math, the notions of measurement, quantity, classification, and so on are the ones that will be used to better understand the meaning of quality of life; in sciences, the notions of equilibrium, space, and energy are some of the concepts offered to better study the focal point; in social studies, teachers will focus, among others, in the notions of space, property, and public good. Curriculum is written without having a fixed content as the starting point, but based on the research conducted in the communities, on the knowledge that the communities offer about the focal point of the thematic complex, and on the scientific knowledge that teachers bring.

The traditional rigid disciplinary structure is broken and general interdisciplinary areas are created. Students will still learn the history of Brazil and the world, "high" culture, and so on, but this will be seen through different lenses. Their culture will not be forgotten in order for them to learn "high status" culture. This means that the Citizen School has embarked on a dual path. It has recognized the necessity of creating empowered channels where people can speak openly, but it also knows that at the same time one must unveil the meanings behind these voices, question their hidden presuppositions, and construct new knowledge. Beginning from the insights of the community, it is necessary not to stop there, but rather to construct knowledge that fights discrimination, racism, and exclusion.

Developing a Critical Discourse of Inclusion, Building Another School

The association between democratic forms of citizenship and schooling is quite old, and countless programs have been implemented to foster this association. It is often the case, in both well established and consolidating democratic regimes, that the goals of those programs go beyond the strictly "pedagogical" and aim not only to maintain but also to deepen democratic processes and forms of governance (Torney-Purta, 2001). Nevertheless, as Myers (2007) note:

> Official curricula often portray democratic citizenship from an elitist perspective in which citizenship means leadership for a few and uncritical patriotism for the many. Textbooks have typically had a limited representation of political participation and of the role of citizens in democracy, while focusing on state authority (Avery and Simmons, 2000–2001; Davies and Issitt, 2005). Large-scale studies measuring students' civic competencies have also focused on formal political

institutions and abstract knowledge of democracy (e.g., Torney-Purta 2001; Weiss 2001). (Myers, 2007, 19)

Myers' critical perspective about the limitations of common forms of citizenship education are similar to the conclusions of Reimers (2007) about the restrictions of a regional program of civic education in Latin America. These two authors offer a good point form of comparison to assess the potential of *Escola Cidadã*.

Supporting and extending real and workable democratic reforms in schools, such as the ones described previously, require that controversial positions be debated fairly within political and social institutions as well as educational systems. The previous examples point to the urgent need for a greater understanding of the sociopolitical conditions facing many aggrieved communities within the social order and the manifold challenges entailed in cultivating civic responsibility to ensure that public institutions embody community decisions with socially acceptable outcomes.

In short, the case of *Escola Cidadã* indicates one road leading toward the horizon of possibility charted out by Freire's ideas. The utopian hope inscribed in this educational experiment is present in its promise to implement democratic experiments within schools (Azevedo, 2007). These experiments are happening on a large scale in Porto Alegre and on a smaller scale in many more instances around the world (Educative Cities Network, Real Utopias Project, Kerala Schools, etc.).

While the dominant trend in educational reform worldwide promotes privatization, high-stakes testing, quantitative indicators of accountability, and fragmented and shallow curricular packages that ultimately blame teachers and students for the lack of relevance of urban public schooling, the public municipal schools of Porto Alegre are an example of a realizable utopia that is based on a critical discourse of hope and inclusion.

The Citizen School Project shows that it is possible to create an alternative space where articulations can be forged and where a new common sense around education can be created. It is possible to create a space where kids and the community feel connected to their schools and feel that their school serves them. Democratizing access, knowledge, and relationships in the current context of global capitalism is not easy, but the fact that *Escola Cidadã* is still operating after decades of structural adjustment in Brazil and has expanded its service to include an increasing proportion of the impoverished communities of greater Porto Alegre reflects the power of a critical discourse of inclusion and community organizing in the struggle for democracy.

NOTES

1. In 2005, after sixteen years in the municipal administration,—the representatives of the Workers' Party were reelected for four consecutive periods (1989–2005)—a centrist alliance won the municipality of Porto Alegre promising not to change the major set of policies implemented by Workers' Party. The structures developed by the *Citizen School Project* are still in place, although the name is no longer associated to the official educational policies of the municipality of Porto Alegre.

2. It is important to note that in the Brazilian educational system, both the states and the cities maintain public schools. In Porto Alegre there are many students who attend state public schools, which explain the relatively small number of students in the municipal schools.

3. The OP also debated citywide matters such as public transportation, health and social assistance, economic development and taxation, urban development, education, culture and leisure.

4. Evelina Dagnino notes that, "Initiated in Porto Alegre, in the south of Brazil, in 1989, participatory-budget experiments exist today in around 100 other cities and are coming to be considered as models for countries such as Mexico, Uruguay, Bolivia, Argentina, Peru, and Ecuador. Because of their success, participatory budgets have recently been adopted by other parties in Brazil, some of them clearly for electoral purposes" (Dagnino, 2003, 7).

5. The cycles were an alternative to the traditional structure of grades with the duration of one year from 1st to 8th. This model was implemented in Belo Horizonte, another Brazilian city governed by a coalition of Leftist parties, and was being implemented in other countries—like Spain and Portugal—and actually would be listed, in the national law, as one of the possible alternatives for school configuration in Brazil. Therefore, what the Citizen School was implementing was not new per se, but a new configuration that, according to the SMED, would offer a substantially better opportunity for dealing with the need for democratization of access and knowledge.

6. For students with special needs, the *Integration and Resources Rooms* were implemented. These "are specially designed spaces to investigate and assist students who have special needs and require complementary and specific pedagogic work for their integration and for overcoming their learning difficulties" (SMED, 1999, 50).

REFERENCES

Abers, R. (1998). "From Clientelism to Cooperation: Local Government, Participatory Policy and Civic Organizing in Porto Alegre, Brazil." *Politics & Society*, 26(4), 511–537.

Anderson, G. L., Herr, K., and Nihlen, A. (1994). *Studying Your Own School: An Educator's Guide to Qualitative Practitioner Research*. Thousand Oaks, CA. Corwin Press.

Anyon, J. (2005). *Radical Possibilities: Public Policy, Urban Education, and a New Social Movement*. New York: Routledge.

Avritzer, L. (1999). "Public Deliberation at the Local Level: Participatory Budgeting in Brazil." Unpublished manuscript. Retrieved from http://www.ssc.wisc.edu/~wright/avritzer.pdf.

Azevedo, J. C. (1998). "Escola Cidadã: Construção Coletiva E Participação Popular." In Silva, L. H. (ed.), *A Escola Cidadã no contexto da globalização*. Petrópolis, Brazil: Editora Vozes. 308–319.

———. (1999a). "Escola Cidadã: Construção Coletiva E Participação Popular." Paper presented at The Comparative and International Education Society, Toronto, April 14–19.

———. (1999b). "A democratização do Estado: A experiencia de Porto Alegre." In Heron da Silva Lui (ed.) Petrópolis, *Escola Cidadã: Teoria e Práctica*. Petrópolis, Brazil: Editora Vozes.

———. (2000). *Escola Cidadã: Desafios, diálogos e travessias*. Petrópolis, Brazil: Editora Vozes.

———. (2007). *Reconversão cultural da escola: mercoescola e escola cidadã*. Porto Alegre, Brazil: SULINA-Editora Meridional.

Baierle, S. (1998). "The Explosion of Experience: The Emergence of A New Ethical-Political Principle in Popular Movements in Porto Alegre, Brazil." In S. Alvarez, E. Daguino, and A. Escobar (eds.), *Culture of Politics, Politics of Culture*. Boulder: Westview Press. 118–138.

Baiocchi, G. (1999). "Participation, Activism, and Politics: The Porto Alegre Experiment and Deliberative Democratic Theory." Unpublished manuscript. Retrieved from http://www.ssc.wisc.edu/~wright/Baiocchi.PDF.

Bowles, S. and Gintis, H. (1986). *Democracy and Capitalism*. New York: Basic Books.

Dagnino, Evelina (2003). "Citizenship in Latin America: An Introduction." *Latin American Perspectives*, 30(2), 3–17.

Fischman, G., Ball, S., and Gvirtz, S. (eds.) (2003). *Crisis and Hope: The Educational Hopscotch of Latin America*. New York: Routledge.

Fischman, G. and McLaren, P. (2005). "Rethinking Critical Pedagogy and The Gramscian Legacy: From Organic to Committed Intellectuals." *Cultural Studies Critical Methodologies*, 5(4), 425–447.

Fischman, G., McLaren, P., Sünker, H., and Lankshear, C. (eds.) (2005). *Critical Theories, Radical Pedagogies and Global Conflicts*. Lanham, MD: Rowman and Littlefield.

Freire, P. (1993). *Pedagogy of the Oppressed* (New rev. 20th-Anniversary ed.). New York: Continuum.

———. (1997a). *Pedagogy of Hope: Reliving the Pedagogy of the Oppressed*. New York: Continuum.

———. (1997b). *Pedagogy of the Heart*. New York: Continuum.

———. (1999). *Pedagogy of Freedom: Ethics, Democracy, and Civic Courage*. Lanham, MD: Rowman and Littlefield.

Gandin, D. (2004). *Planejamento como prática educativa*. Petrópolis, Brazil: Vozes.

Genro, T. (1999). Cidadania, emancipação e cidade. In Silva, L. H. (ed.), *Escola Cidadã: Teoria e prática*. Petrópolis, Brazil: Vozes. 7–11.

Giroux, Henry and Peter McLaren (1997). "Paulo Freire, Postmodernism, and the Utopian Imagination: A Blochian Reading." In Jamie Owen Daniel and Tom Moylan (eds), *Not Yet: Reconsidering Ernst Bloch*. London and New York: Verso. 138–162.

Hursh, D. (2006). "The Crisis in Urban Education: Resisting Neoliberal Policies and Forging Democratic Possibilities." *Educational Researcher*, 35(4), 19–25.

Klees, Steven J. (2006). "A Quarter-Century Of Neoliberal Thinking In Education: Misleading Analyses And Failed Policies." Presented at the World Bank conference on *The Contributions of Economics to the Challenges Faced by Education*. University of Dijon, France, June 21–23, 2006. 1–49. Retrieved from http://www.u-bourgogne.fr/colloque-iredu/poster-scom/communications/Pa83StevenKlees.pdf.

Krug, A. (2001). Ciclos de formação: uma proposta transformadora. Porto Alegre: Mediação.

Myers, John P. (2007). "Citizenship Education Practices of Politically Active Teachers in Porto Alegre, Brazil and Toronto." *Canada Comparative Education Review*, 51(1), 1–22.

Ozga, J. (2000). "Education Policy in the United Kingdom: The Dialectic of Globalisation and Identity." *Australian Educational Researcher*, 27(2), 87–97.

Reimers, F. (2006). "The Public Purposes of Schools in an Age of Globalization." *Prospects*, 36(3), (September), 1–24.

Santos, B. S. (1998). "Participatory Budgeting in Porto Alegre: Toward a Distributive Democracy." *Politics and Society*, 26(4), 461–510.

Schugurensky, D. (2001). "Grassroots Democracy: The Participatory Budget of Porto Alegre." *Canadian Dimension*, 35(1), 30–32. Jan/Feb.

SMED. (1999). "Ciclos de formação—Proposta político-pedagógica da Escola Cidadã." *Cadernos Pedagogicos*, 9(1).

SMED. (2000). *Boletim Informativo—Informações Educacionais*. Year 3, No. 7.

Shepard, L. A. and Smith, M. L. (1989). "Flunking Grades: A Recapitulation." In Lorrie A. Shepard and Mary Lee Smith (eds.), *Flunking Grades: Research and Policies on Retention*. London: Falmer Press. 214–235.

Torney-Purta, Judith. (2001). *Citizenship and Education in Twenty-eight Countries: Civic Knowledge and Engagement at Age Fourteen*. Amsterdam: IEA.

Zeichner, K. and Noffke, S. (2001). "Practitioner Research." In V. Richardson (ed.), *Handbook of Research on Teaching*, 4, 298–330. New York: American Educational Research Association/Macmillan.

CHAPTER 5

Persistent Exclusions in Postapartheid Education: Experiences of Black Parents

Bekisizwe S. Ndimande

Pedagogies of exclusion in education were officially and formally implemented in South Africa as early as 1658 when the first slave school was set up by the Dutch East India Company to teach good values to the indigenous population, as well as the enslaved peoples brought to the Cape Colony, to work in the plantations and to become faithful servants to the masters.[1] These pedagogies of exclusions were further implemented by the British missionary school system whose aim was to "civilize" and "Christianize" the indigenous population. By 1953, a more racially exclusive system was imposed by apartheid to deny and marginalize the black citizens access to education (Kallaway, 1984; Christie, 1985). Put differently, black citizens in that nation were excluded through an inferior education system called Bantu education, which offered "pedagogies of oppression" that enforced social subordination and the ideology of white supremacy.[2]

The victory over apartheid in 1994 was followed by a democratic Constitution (1996) and the passing of the South African Schools Act (SASA, 1996), whose purpose was to repeal all apartheid schooling legislation and put in place a democratic curriculum and policy that sought to institute a uniform, nonracial system in terms of school organization, governance, funding, and other institutional aspects in order to establish inclusive education. The policies toward education inclusion were regarded by many as an affirming project that could harness social cohesion through education equality in a nation that excluded the subaltern groups in every social sphere for many decades. Yet a close examination in many postapartheid public schools shows some lingering pedagogies of exclusion despite all the effort to transform education so that it fulfills the promise for inclusive education.

This chapter provides a critical examination of the educational transition from the "pedagogies of oppression" under apartheid toward policies for educational inclusion in the postapartheid period. I argue that some of the crucial changes that the postapartheid government initiated with the intent for inclusive education have been immensely challenged and compromised by the hegemonic responses in those schools and educational institutions that resist inclusive education initiatives. First, I present a brief history of education from apartheid to the transition period—post-1994—to provide a contextual background to this analysis. Second, I foreground the black parents' voices to examine the financial and racial barriers facing education inclusion in South Africa. These voices help us better understand the challenges to the more inclusive postapartheid education policies.

Third, I argue that the inadequate funding in most township schools, as well as the racial exclusions in most suburban schools, constitutes the undesirable character of education exclusion in a nation that has endorsed the implementation of inclusive policies in its institutions. Drawing on critical race theory, I explain why it is crucial for postapartheid to vigorously address the problems of racism by insisting on inclusive education as opposed to the persistent exclusionary education policies that have effectively marginalized the majority people in that nation. I argue that inclusive education policies can help build social cohesion and thus solidify the democratic values of social inclusion enshrined in the democratic Constitution in order to uproot the exclusionary discourses that underprivileged the subaltern communities of that country for many decades.

HISTORICAL ANALYSIS

Bantu education was introduced by H. F. Verwoerd, the Minister of Native Affairs in 1953, with the support of the white-only parliament of the time. It was the most brutal type of educational discrimination that was enforced on black students in the history of South Africa (Kallaway, 1984; Christie, 1985; Nkomo, 1990).[3] Besides its discriminative character, Bantu education also enforced an inferior and hegemonic curriculum policy on segregated black public schools. For Verwoerd, all schools had to adhere to their assigned functions, and black public schools in this case were required by law to teach Bantu education, which sought to impose inferiority on teachers and students with the sole requirement of bureaucratic and political compliance (Jansen, 2001).

This was a case in which education functioned as the ideological state apparatus (Althusser, 1971) to socially, economically, and politically exclude the black communities by subjugating them into subservient positions. Louis Althusser argues that the state has two forms in which it exerts control over its people, namely, the repressive state apparatuses (RSA) and the ideological state apparatuses (ISA). According to this analysis, it is the latter that is associated with the indoctrination of the mind, in this case, Bantu education. While Althusser makes an important claim, it is also true that such analysis requires a deeper inquiry that examines agency and subjectivity as discursive entities. Others have argued about the limitations of "correspondence theories" by pointing out the power of agency, basically arguing that people are not dupes who are readily interpolated into dominant ideology without realizing its adverse effects, or that they lack the abilities to subvert such ideologies (Willis, 1977; Cole, 1988; Apple, 1996). While acknowledging some of the realities in correspondence analyses, this body of research simply warns against essentializing arguments that tend to view tacit strategies and ideologies as always guaranteed for intended outcomes.[4]

The demise of apartheid brought along a whole array of democratic changes in a nation ravaged by policies of racial exclusion for more than four decades. With a progressive agenda, the postapartheid state adopted a democratic Constitution in May 1996, immediately followed by the South African Schools Act (SASA), whose principal goals were to install democracy and eradicate racial and educational exclusions. As pointed out by Samoff (2001), education had an important role in the demise of apartheid as well as in the transformation toward inclusive schooling. While Samoff's remarks are crucial, Wong and Apple (2003) warn us to be cautious and to not exaggerate the role of education as if it is a dependent variable influenced by the state policies. They remind us about the subtle dynamics about state formation and the role of education, which means that education *can* also work to counter the intended goals of the state, and in this particular case undermine the state's vision of inclusive policies in education.

As we can see in this context, despite the firm commitment by the postapartheid government to promote inclusive education for all students, township schools ironically remained entirely racially segregated and under-resourced (Vally and Dalamba, 1999; Chisholm, 2004; Ndimande, 2006; Jansen and Amsterdam, 2006). The lack of adequate resources has left these public schools dysfunctional and with poor results, which effectively left children from these schools underemployed after graduation and with less chances of access to college education.[5] But this was not the only challenge to inclusive education. Most of the former white-only schools in the suburban areas have access to better resources, and these schools were desegregated to allow students from poor schools access, yet they continued to practice racial exclusion against black students who enrolled in these schools (Vally and Dalamba, 1999; Jansen, 2004; Soudien, 2004; Ndimande, 2005). It was soon clear that these challenges undermined the democratic goals of education for all (Brock-Utne, 2000) and the initiatives toward pedagogies of inclusion in postapartheid South Africa.

As I show in this discussion, these challenges mainly affected the subaltern groups, in particular the black communities. But this shouldn't be surprising because social exclusion under apartheid was mainly based on race in South Africa, although the problems of social class stratification, gender oppression, cultural and (dis)ability exclusions were also notable. Vestergaard (2001) reminds us that the cornerstone of apartheid was racial discrimination and the institutionalization of white supremacy, thus all communities who were not white were excluded in this apartheid configuration.[6]

With regard to education, this is how Nkomo et al. (2004) put it: "The role of race in promoting racial division was seen as central. Whether linked to class or to culture and language, race and racism were inextricably part of the social fabric and fundamentally shored up by the education system and its schools" (5). Therefore, my arguments about the challenges facing the pedagogies of inclusion in postapartheid South Africa are largely situated within the politics of race, since it is the black and poor who are on the receiving end of this continuum. I particularly focus on the voices of the marginalized groups as a case study in this discussion to help us better understand the challenges to the more inclusive educational policies in the twenty-first-century South Africa.

Parents Voices on Resource Exclusions

At this point I would like to introduce two black parents, Mapule and Seipati, who talk about their specific experiences in public schools with regard to the pedagogies of inclusion.[7] As they shared their experiences and observations, Mapule and Seipati were concerned about the resource situation in township schools where one of the parents sends their children.[8] But first, let us hear from Mapule, a middle class black woman who lives in one of the black townships, southeast of Johannesburg, in the Gauteng province. Mapule is one of the parents who believed that education is important in order to succeed in this age of global economy. This is how she described her observations and experiences about the public schools in the townships and those that serve predominantly white students in the suburbs where her own children attend.

> Mapule: There are no facilities in our black schools. I discovered that [formerly][9] white [only] schools have all the resources, for example, natural science experiment materials. In our [black] schools, when you speak of a glass beaker, you have to draw it on the board because there aren't any materials for experiments. Children won't be able to see the real one. But in [formerly] white schools, they do have those materials. In short, I took my children there for better facilities. (Trans. KM FG 6/5/03)

Like Mapule, Seipati's family lives in the same township and she holds similar views regarding the lack of resources in township schools. Unlike Mapule who is middle class and has been able to transfer her son and daughter to a suburban school that has adequate resources, Seipati is from the working class and unlikely to have sufficient financial means to "escape" poor township schools, because of her material conditions.

> Seipati: We [in township schools] don't have resources—we don't have computers. We need resources so that we don't have to wish to send our children to formerly white-only schools. (Trans. PCD, 3/15/03)

Mapule and Seipati's concerns about resources in township schools are shared by the other parents who participated in the study. These are expressions by parents on what they perceive as persistent exclusions in public schools through unequal access to resources between segregated township schools and desegregated former white-only schools in the suburbs. Motala (2006) reminds us that while the resource discrimination has been removed, the "historical backlogs have been difficult to redress and the gains of increased expenditure have been eroded by inflation. Private expenditure in the form of fund-raising and school fees changed the picture of equalisation to one of substantial differentiation within the public schooling sector" (80).

Apart from the historical backlog, the problem of inadequate resources is also exacerbated by under-spending by the state. Clearly, the postapartheid government has committed to funding education (Jansen, 2005; Department of National Treasury, 2008). However, a substantial amount of money budgeted for improving the country's most disadvantaged public schools get trapped in the government bureaucracy and barely reaches those schools with urgent needs for resources, thus remaining unspent at the end of the fiscal year (MacFarlane, 2002).[10] Sayed (2001) and others argue that this phenomenon of under-spending is also related to postapartheid government officials' inability to understand and deal with the bureaucracy.

Yet it is important to realize that this kind of under-spending is neither neutral nor apolitical. It is related to a broader neoliberal ideology, that of bashing public institutions as being inefficient and wasteful; this is basically an ideological strategy of reducing government control of economy and the reduction of public spending. For neoliberals, as pointed out by Apple (2001), "Public institutions such as schools are 'black holes' into which money is poured—and then seemingly disappears—but which do not provide anywhere near adequate results" (38). Such ideologies come with lots of pressure to institutions supported by public funds, which at times calls for reduction of support for the common good. But reduction and under-spending has a negative impact because it tends to protect the interests of the wealthy class and neglect the poor; as evident in this context where most of the affected schools are those located in black and poor neighborhoods.

The main agenda behind neoliberalism is the privatization and marketization of the public sphere so that individuals must compete for their own social mobility and success (Chomski, 1999; Lauder and Hughes, 1999; Apple, 2001). In this political discourse, individuals will be rewarded according to their ability (Gillborn and Youdell, 2000) to compete in the "free and neutral" terrain called the market. But this line of thought is problematic in that it doesn't consider the social field of power in which this competition takes place. Nor does it recognize the previous social exclusions to which the marginalized groups have been subjected.

How is it possible that a nation such as South Africa, just emerging from apartheid, gets associated with neoliberal ideologies of social exclusion? Bond (2005) argues that even before the dismantling of apartheid, the South African economic landscape had drastically shifted from what he refers to as a popular-nationalist, antiapartheid project, toward the global economic framework largely influenced by the World Bank and the International Monetary Fund. This economic shift has subsequently influenced the country's public policy toward neoliberalism (Desai, 2002; Garson, 2002; Pillay, 2002; Monbiot, 2004; Pampallis, 2004; Bond, 2005; Gumede, 2005).[11] Pampallis (2004), for instance, points to the White Paper of 1987, which allows the government to engage external educational agencies to under-take tasks previously done by the National Education Department. This included contracts to consultants with multinational agencies in core education functions. Therefore the under-spending of educa-tion budget, when it can help previously excluded schools, should be viewed within this public/private nexus.

Apart from education exclusions based on resources in township schools, black students have also been racially excluded in desegre-gated schools. Since township schools are historically under-resourced and have been inadequately funded in postapartheid, black parents started to transfer their children to adequately funded former white-only schools in the suburban areas.[12] This created a scenario in which school desegregation became a one-sided process, with black children transferring to former white-only schools in the suburban areas, but not the other way round.[13] However, this migration was not going to address the needs for better education for these children. Instead, black children were faced with a different kind of challenge—namely, racism. A groundbreaking study on school integration in postapart-heid (Vally and Dalamba, 1999) found that the race relations in these schools raised serious concerns. For example, this report found that there was a notable evidence of racial discrimination and racial inci-dents among students, including racial slurs directed at black stu-dents. Subsequent analyses (Jansen, 2001, 2004; Nkomo et al., 2004; Ndimande, 2005) draw similar conclusions about the troubling race relations in former white-only schools.

PARENTS VOICES ON RACIAL EXCLUSIONS

Let me introduce NomaSwazi and Nobelungu, two black parents who live in different neighborhoods—a black township and a sub-urban area in the Gauteng province, but both have transferred their children to former white-only schools in suburban areas. These two parents talk about their children's experiences in these schools. Let me begin with NomaSwazi, who is a middle-class woman by vir-tue of her profession as a primary school teacher in the township of Mhlenge, east of Johannesburg, where she and her family live.

NomaSwazi told me that she knows firsthand about the resource situation in township schools, hence she decided to transfer her daughter to a public school in a nearby suburban area. In the interview, she mentioned several positive things about the school, for example, the abundance of resources, access to technologies, learning to speak and write English fluently, and so forth (Ndimande, 2005). Here is part of that conversation:

Question:
What are some of the limitations or bad things that trouble you about formerly white-only schools in the suburban areas?

NomaSwazi:
My child complains about racism. While they admit black students, our children are not treated the same as white children. My son complains about one of their head teachers who makes remarks such as: "You coloured boy" or "You black girl" when referring to our children. I think this is problematic...I once called the principal up and complained about this. She told me that she would attend to it...You see, if a coloured or a black child did something wrong, that is when you hear the noise, but there won't be any [noise] if a white child did a wrong thing. (Trans Mono FG. 4/8/03)

A similar observation is related by Nobelungu, also a middle-class woman who, unlike NomaSwazi, lives in the suburban area of Mthonjane. Though her family has since moved to live in the suburban area, she remains as a school teacher in a township school where she commutes daily. Like NomaSwazi, Nobelungu has enrolled her son in a former white-only public school in the suburb where the family now resides. This is how she responds to the interview question:

Question:
What are some of the limitations or bad things that trouble you about formerly white-only schools in the suburban areas?

Nobelungu:
Let me cite an example of racism in these schools. There is this school, I won't mention its name...[T]hey used to call parents' meetings. However, the parents' meetings were called on different days for different parents. The principal would call a meeting for black parents today and the next day another meeting for white parents. So it was just segregated. There was not even a single complaint from the School Governing Body (SGB)...[Y]ou know, even if you went to the principal and talked to him about this, he would just tell you that it has been like this from the beginning. (Trans. IJ, 3/17/03)

The voices of these parents express disquieting race relations in these schools, a complicated learning environment that has shown evidence of persistent practices of exclusion through education. Apart from these voices, Jansen's (2001) analysis of schools in KwaZulu Natal, particularly at Westville, shows how some white parents actively try to block the desegregation of white schools by using arbitrary zoning policies, thus excluding black children from schools with resources.

Yet there are other practices that have caused persistent exclusions in the postapartheid education. In the years 1995 through 1998, just after the democratic elections of 1994, a racially charged conflict happened at Vryburg High School, a predominantly Afrikaans-speaking public school in a small white conservative rural town of Vryburg, in the North West province. The conflict started when black students from Huhudi, a nearby black township, had to be integrated to Vryburg High School because of overcrowding problem in their township schools (Odhav, Semuli, and Ndandini, 1999). Upon integration, the school principal, with the help of the predominantly white School Governing Body, arbitrarily used the school's language policy to exclude black students.[14] In addition, the school increased fees in what appeared to be a proxy for access denial to poor black students. Vryburg was in the national news following this divisive racial clash at their school. This issue eventually went to court, where the Constitutional Court subsequently ruled in favor of the white principal. According to Jansen (2004), this is a specific case where conservative Afrikaner parents in this predominantly white public school sided with their white offspring in trying to deny access to black students. And how many other "Vryburg" cases happens, asks Jansen, that people do not get to hear or read about.

These challenges experienced by black parents in former white-only schools do not, by any measure, support the initiatives toward inclusive education. Given the parents' voices represented in this discussion, it is therefore arguable that inadequate funding and racism has become one of the barriers to policies of inclusive education in postapartheid schools. In a very paradoxical way, the racialization of the school zones by conservative white parents, in spite of the South African Schools Act (1996), has also foregrounded white privilege and its exacerbation of social and educational exclusions between and among communities of different races.

Race as a social category is complex, unstable, and can be constructed or de/reconstructed in contradictory ways. Because of this complexity in the construction of race, it becomes imperative that it be understood at a theoretical level as well, especially because of its influential role in the persistent policies and practices of exclusion in this context. Just because a country has instituted policies of social inclusion doesn't imply that it will necessarily translate to successful implementation of these progressive policies, especially when racial inequalities continue to be a threat. Therefore, as I argue below, race as a social construct needs to be understood within its discursive manifestation in these contexts.

UNDERSTANDING PEDAGOGIES OF EXCLUSION THROUGH CRITICAL RACE THEORY

My analysis in this chapter is framed within the discourse of critical race theory. Omi and Winant (1994), Ladson-Billings and Tate (1995), Dyson (1996), Gillborn and Youdell (2000), Delgado and Stefancic (2001), Pollock (2006), and many others have made scholarly contribution to debates around race, racism, and racial injustices in our schools and other social institutions. This corpus of literature reminds us about the constitutive role the power of race plays in societies emerging from or dealing with racial inequalities. Ladson-Billings and Tate (1995) contend that race is crucial to the examination of issues of diversity and inclusion policies in schools. They argue that race continues to be significant in explaining inequity in education, and that class and gender alone are not sufficient categories of analysis to explain all the difference in school experiences. Of course, we cannot afford to reduce race relations to a mere manifestation of other supposedly more fundamental social and political relationships such as class and ethnicity (Omi and Winant, 1994). When most patterns of racial relations reveal themselves to be more complex and contradictory trajectories (Omi and Winant, 1994), we cannot afford to think of our schools' policies and social system as color-blind in a racialized nation like South Africa.

Race has plagued education reform in many contexts, an experience that has been criticized by Gillborn and Youdell (2000), whose research found that education reform in Britain focused mainly on such arbitrary things as "ability," tests scores, standards, and parental choice, as if these were the answers to pedagogies of exclusions. In fact, argues Gillborn and Youdell, all these neoliberal and neoconservative reforms ignore the real diversity of experience and social justice issues and actually serve to remove social inclusion policy and practice from the school reform agenda, allowing racial inequalities to go unnoticed and unchecked. Again, we can see here the power of race at institutional levels, which often privileges the dominant groups and marginalizes those from the minority backgrounds, thus perpetuating pedagogies of exclusion.

CONCLUSION

Since emerging from socially excluding policies of apartheid, it was critical for postapartheid South Africa to implement policies of inclusion in schools to undo the inequalities of the apartheid. Therefore, it was appropriate and necessary for the postapartheid institutions to implement multiple democratic projects aimed at transforming the nation toward social inclusion in all these institutions. The educational system became the major focus of this transformation program since education is viewed as a major tool of social transformation. As noted by Freire (2007), it is education itself that needs priority in transformation because a transformed education system can help challenge those who oppress others, since it is democratic education that is often denied to a large part of a population.

As I have argued in this chapter, while there has been a concerted effort at social inclusion through schools, especially through increasing funding of public schools of that nation, the bottom line remains that most township schools have excluded poor children because of inadequate funding in these schools. At the surface, the problems seem to be related to mismanagement and bureaucracy, yet a critical analysis shows that the problems of access and funding are much more complex. I have, therefore, framed this problem within the broader neoliberal ideology that the South African economic policy has adopted. I have also shown that racism becomes a barrier to township children when they transfer from poor schools in their neighborhoods to former white-only schools in the suburbs in search of better educational resources. Therefore, financial exclusion of township children get even worse by the racial exclusions these children encounter in suburban schools.

We read the compelling testimonies from Mapule, Seipati, NomaSwazi, and Nobelungu, all black parents who explicitly showed that the principle of inclusive education has been compromised by inadequate funding in township schools and by the excluding policies in better-funded schools. Undoubtedly, most of these parents, especially the poor, have been denied the promise of inclusive education for equal opportunities. In addition, black students who experience racism as described by their parents are adversely affected to an extent that their educational opportunities are compromised. As I pointed out earlier, this does not mean to suggest that the postapartheid government has backpedaled on social inclusion philosophy. Rather, I am concerned that without rigorous monitoring of practice in desegregated schools and without seriously evaluating the impact of the economic policies and the influence of neoliberal and neoconservative impulses, social exclusions may be (re)produced through the education system.

The absent presence of race in these schools can have a consequential effect on black students' educational outcomes and their future opportunities. Nations that have been grappling with policies of inclusion for decades are still hindered by the problem of race, in addition to the challenges of the influence of neoliberalism, which makes it pertinent for scholars and researchers to understand race and the politics of neoliberalism in the framing of current reform initiatives for social inclusion policies. South Africa has just emerged from apartheid fourteen years ago, and here too neoliberalism and racism seem to be an overarching hindrance to inclusive education.

I concur with Nieto (2006) that in order to implement successful pedagogies of inclusion, we need to eliminate what she calls the "resource gap" in our public schools. Nieto maintains that we should provide students with all resources necessary to learn to their potential. Indeed, as Nieto further argues, working toward democracy also means adopting a conscientious decision to end social poverty by "opening the doors" of learning to all students, a mantra that was dominant during the People's Education movement in the 1980s.[15] Most importantly, all stake holders should interrogate any forms of systematic exclusions in public schools of this nation that hinders its democratic progress. Public schools in democratic nations are largely funded by taxpayers' money and it only makes sense that no children should be denied quality education due to the lack of resources in schools in their own neighborhoods, or because of conservative segments of white communities who deny access to black children in public schools with resources. Simply put, township schools should have access to adequate resources and black children should not be denied access to schools with better resources outside their neighborhoods simply because of the color of their skin. Persistent pedagogies of exclusion in postapartheid must be vigorously interrogated to make way for inclusive education in order to realize the dream of liberation from apartheid.[16]

Notes

1. Molteno's analysis on slave schools in the Cape Colony is absolutely essential for anyone interested in the history of the pedagogies of exclusions and the origins of school segregation in South Africa.

2. See Ndimande (2005), especially on p. 105, for a description of an ordinary day in school for black students who experienced educational exclusion in Bantu education.

3. As I show in the beginning of this chapter, Bantu education as an oppressive system was preceded by other forms of hegemonic education systems, yet I focus on Bantu education since it has the long-standing impact on the pedagogies of exclusions in South African history of education.

4. In South Africa, for instance, instead of ISA (Bantu education) indoctrinating people to become docile recipients of apartheid, massive resistance erupted that would eventually lead to the demise of the hegemonic rule in 1994. See Motlhabi (1984) who analyzes black resistance to apartheid.

5. Haroon Bhorat's (2004) study of labor market and unemployment trends in postapartheid South Africa shows the continued increase of unemployment amongst poorly educated children. The number is high (47 percent) among black South Africans and low (9 percent) among white South Africans.

6. Apartheid created four racial categories in that nation: blacks, coloureds, Indians, and whites. All but the whites were subjected to extreme social exclusion. Again, Motlhabi (1985) is helpful in this discussion.

7. The data come from the study that was conducted to evaluate the common sense guiding black parents regarding school "choice" in postapartheid South Africa, see Ndimande (2005). *Cows and goats no longer count as inheritances.*

8. These and all other subsequent names of people and places in this article are fictitious.

9. It is worth mentioning that Mapule refers to the recently desegregated schools in the suburbs as "white schools." Using such a language is not a simple misunderstanding of the recent changes (that schools have now desegregated). Rather, this suggests a symbolic language to indicate the unequal socioeconomic landscapes regarding who has access to which kind of school. It also implies that although they are allowed to send their children in these schools, black parents do not perceive themselves as "co-owners" because of the attitudes and the manner in which these schools relate to them.

10. Previously, money had been returned to the Department of National Treasury because it was under-spent or never used during the fiscal year when it was allocated to reach and help increase resources in poor schools.

11. Bond (2005), in particular, provides a historical and insightful analysis of these economic policy changes from apartheid to democracy, especially the introductory chapter, *Dissecting South Africa's Transition.*

12. See Vally and Dalamba (1999); Ndimande, 2005; and Soudien, (2004). Unlike Vally and Dalamba (1999) and Ndimande, 2005, Soudien (2004) shows a different dimension where migration from

township schools leads toward predominantly Indian and coloured public schools.

13. This is not a unique South African experience. A similar pattern of school migration took place after the Supreme Court had ruled school segregation illegal in the United States in 1964. Here too it was African American children, not European American, who were bused to white desegregated schools outside African American neighborhoods. Orfield (2004) states that after the Supreme Court ruling, there was a dramatic increase of black children in predominantly white schools. See also Clotfelter (2004).

14. At first, black students were placed in separate classrooms where they were taught in English while white students were in the other classrooms where they were taught in Afrikaans. Eventually the school decided to combine classes and enforced Afrikaans as a medium of instruction (see Odhav, Semuli, and Ndandini, 1999). Afrikaans as an enforced medium of instruction had previously led to protests and riots in black schools—the 1976 uprisings in Soweto and throughout the country—which led to broad mobilization against apartheid.

15. People's Education was a social movement that originated from the antiapartheid struggle in the 1980s to demand the abolition of apartheid education and the implementation of inclusive education for all (see Motala and Vally, 2002).

16. I would like to thank the editors of this volume for their valuable feedback that immensely shaped the arguments in this chapter, especially Beth Blue Swadener who provided helpful suggestions in various stages of this piece. My thanks also go to the parents who shared with me their lived experiences in postapartheid education.

REFERENCES

Althusser, L. (1971). *Lenin and Philosophy and Other Essays*. New York: Monthly Review Press.

Apple, M. W. (1996). *Cultural Politics and Education*. New York: Routledge.

———. (2001). *Educating the "right" Way: Markets, Standards, God, and Inequalities*. New York: Routledge.

Bhorat, H. (2004). "The Development Challenge in Post-Apartheid South African Education." In L. Chisholm (ed.), *Changing Class: Education and Social Change in Post-Apartheid South Africa*. Pretoria: HSRC Press. 31–55.

Bond, P. (2005). *Elite Transition: From Apartheid to Neoliberalism in South Africa* (2nd ed.). Pietermaritzburg: University of KwaZulu Natal Press.

Brock-Utne, B. (2000). *Whose Education for All?: The Recolonization of the African Mind*. New York: Falmer Press.

Chisholm, L. (2004). "Introduction." In L. Chisholm (ed.), *Changing Class: Education and Social Change in Post-Apartheid South Africa*. Pretoria: HSRC Press. 1–28.

Chomski, N. (1999). *Profit over People: Neoliberalism and Global Order.* New York: Seven Stories Press.

Christie, P. (1985). *The Right to Learn: The Struggle for Education in South Africa.* Braamfontein, Johannesburg: Ravan Press.

Clotfelter, C. T. (2004). *After Brown: The Rise and Retreat of School Desegregation.* Princeton, NJ: Princeton University Press.

Cole, M. (ed.) (1988). *Bowles and Gintis Revisited.* New York: Falmer Press.

Constitution of the Republic of South Africa (1996). (As adopted on May 8, 1996, and amended on October 11).

Delgado, R. D. and Stefancic, J. (2001). *Critical Race Theory: An Introduction.* New York: New York University Press.

Department of National Treasury of the Republic of South Africa at http://www.treasury.gov.za/ Accessed March 13, 2008.

Desai, A. (2002). *We are the Poors: Community Struggles in Post-Apartheid South Africa.* New York: Monthly Review Press.

Dyson, M. E. (1996). *Race Rules: Navigating the Color Line.* Reading, MA: Addison-Wesley Publishing Company, Inc.

Freire, P. (2007). *Daring to Dream: Toward A Pedagogy of the Unfinished.* Boulder: Paradigm.

Garson, B. (2002). "Johannesburg and New Jersey Water." [Online]. Available at http://www.nyenvirolaw.org/pdf/Garson-newjerseyandjohannesburg water.pdf. Accessed December 2002.

Gillborn, D. and Youdell, D. (2000). *Rationing Education: Policy, Practice, and Equity.* Buckingham: Open University Press.

Gumede, W. M. (2005). *Thabo Mbeki and the Battle for the Soul of the ANC.* Cape Town: Zebra Press.

Jansen, J. D. (2001). "Access and Values in Education." A paper presented at the "*Saamtrek*" *values, education, and democracy in the 21st century*, Kirstenbosch, Cape Town.

———. (2004). "Race and Education after Ten Years." *Perspectives in Education*, 22(4), 117–128.

———. (2005). "Does Money Matter? Towards An Explanation for the Relationship between Spending and Performance in Education in South Africa." A paper presented at the *Institute for Justice and Reconciliation*, Cape Town.

Jansen, J. D. and Amsterdam, C. (2006). "The Status of Education Finance in South Africa." *Perspectives in Education*, 24(2), vii–xvi.

Kallaway, P. (ed.) (1984). *Apartheid and Education: The Education of Black South Africans.* Johannesburg: Ravan Press.

Ladson-Billings, G. and Tate, W. F. (1995). "Toward A Critical Race Theory of Education." *Teachers College Record*, 97(1), 47–68.

Lauder, H. and Hughes, D. (1999). *Trading in Futures: Why Markets in Education Don't Work.* Philadelphia: Open University Press.

MacFarlane, D. (2002). "Education's Marginalized Millions." *Mail & Guardian*, 18(15), 6. April 19–25.

Molteno, F. (1984). "The Historical Foundations of the Schooling of Black South Africans." In Kallaway, P. (ed.), *Apartheid and Education: The Education of Black South Africans*. Johannesburg: Ravan. 45–107.

Monbiot, G. (2004, October 19). "Exploitation on Tap: Why is Britain Using Its Money to Persuade South Africa to Privatize Its Public Services?" Available at http://environment.guardian.co.uk/water/story/0,,1845319,00.html. Accessed April 9, 2007.

Motala, S. (2006). "Education Resourcing in Post-Apartheid South Africa: The Impact of Finance Equity Reforms in Public Schooling." *Perspectives in Education*, 24(2), 79–93.

Motala, S. and Vally, S. (2002). "People's Education: From peoples' Power to Tirisano." In Kallaway, P. (ed.), *The History of Education under Apartheid 1948–1994: The Doors of Learning and Culture shall be Opened*. Cape Town: Maskew Miller Longman. 174–194.

Motlhabi, M. (1985). *The Theory and Practice of Black Resistance to Apartheid: A Social-Ethical Analysis*. Johannesburg: Skotaville.

Ndimande, B. S. (2005). "Cows and Goats no Longer Count as Inheritances: The Politics of School 'Choice' in Post-Apartheid South Africa." Unpublished doctoral dissertation (Ph.D.), University of Wisconsin-Madison.

———. (2006). "Parental 'Choice:' The Liberty Principle in Education Finance." *Perspectives in Education*, 24(2), 143–156.

Nieto, S. (2006). "Creating New Visions for Teacher Education: Educating for Solidarity, Courage, and Heart." A paper presented at the *American Association of Colleges for Teacher Education* (AACTE), January 30, San Diego, CA.

Nkomo, M. (1990). "Introduction." In M. Nkomo (ed.), *Pedagogy of Domination: Toward A Democratic Education in South Africa*. Trenton, New Jersey: Africa World Press, Inc. 1–15.

Nkomo, M., Chisholm, L., and McKinney, C. (2004). "Through the Eye of the School in Pursuit of Social Integration." In Nkomo, M., Chisholm, L., and McKinney, C. (eds.), *Reflections on School Integration*. Cape Town, Human Science Research Council. 1–10.

Odhav, K., Semuli, K., and Ndandini, M. (1999). "Vryburg Revisited: Education, Politics and the Law after Apartheid." *Perspectives in Education*, 18(2), 43–60.

Omi, M. and Winant, H. (1994). *Racial Formation in the United States: From the 1960s to the 1990s*. (2nd ed.). New York: Routledge.

Orfield, G. (2004). "The American Experience: Desegregation, Integration, Resegregation." In Nkomo, M., Chisholm, L., and McKinney, C. (eds.), *Reflections on School Integration*. Cape Town: Human Science Research Council. 95–124.

Pampallis, J. (2004). "The Education Business: Private Contractors in Public Education." In L. Chisholm (ed.), *Changing Class: Education and Social Change in Post-Apartheid South Africa*. Pretoria: HSRC Press. 421–433.

Pillay, D. (2002). "Between the Market and A Hard Place." *Sunday Times*, October 6, 24.

Pollock, M. (2006). *Colormute: Race Talk Dilemmas in an American School.* Princeton, NJ: Princeton University Press.

Samoff, J. (2001). "'Education for All' in Africa but Education Systems That Serve Few Well." *Perspectives in Education*, 19(1), 5–28.

Sayed, Y. (2001). "Post-Apartheid Educational Transformation: Policy Concerns and Approaches." In Y. Sayed and J. D. Jansen (eds.), *Implementing Education Policies: The South African Experience.* Cape Town: University of Cape Town Press. 250–270.

Soudien, C. (2004). "'Constituting the Class': An Analysis of the Process of 'Integration' in South African Schools." In L. Chisholm (ed.), *Changing Class: Education and Social Change in Post-Apartheid South Africa.* Pretoria: HSRC Press. 89–114.

South African Schools Act, (1996). *Government Gazette of the Republic of South Africa*, NO. 84 of 1996, vol 377, November 15.

Vally, S. and Dalamba, Y. (1999). *Racism, "Racial Integration" and Desegregation in South African Public Secondary Schools.* Pretoria: South African Human Rights Commission.

Vestergaard, M. (2001). "Who Got the Map? The Negotiation of Afrikaner Identities in Post-Apartheid South Africa." *Daedalus: Journal of the American Academy of Arts and Sciences*, 130(1), 19–44.

Willis, P. (1977). *Learning to Labor: How Working Class Kids Get Working Class Jobs.* New York: Columbia University Press.

Wong, Ting-Hong and Apple, M. W. (2003). "Rethinking the Education—State Formation Connection: The State, Cultural Struggles, and Changing the School." In M. W. Apple (ed.), *The State and the Politics of Knowledge.* New York: Routledge. 81–107.

PART II

Curricular Conversations:
Successful Programs and Initiatives

CHAPTER 6

The Curriculum and the Politics of Inclusion and Exclusion

Wayne W. Au and Michael W. Apple

Two plaques marking the death of a well-known European sailor can be found on the Philippine island of Mactan. One of these plaques reads:

> On this spot Ferdinand Magellan died on April 27, 1521, wounded in an encounter with the soldiers of Lapulapu, chief of Mactan Island. One of Magellan's ships, the *Victoria*...sailed from Cebu on May 1, 1521, and anchored at San Lucar de Barrameda on September 6, 1522, thus completing the first circumnavigation of the earth. (Pinguel, Wei, and Shalom, 1998, 15)

In sharp contrast, a second plaque on Mactan, which also commemorates Magellan's death, reads:

> Here, on 27 April 1521, Lapulapu and his men repulsed the Spanish invaders, killing their leader, Ferdinand Magellan. Thus, Lapulapu became the first Filipino to have repelled European aggression. (Pinguel et al., 1998, 15)

The difference in perspective of these historical markers is obvious and simple, but it becomes important if we take their individual historical contexts into account. The first plaque, which takes on the perspective of Magellan and the Spanish colonizers of the Philippines, was erected in 1941. The second, which was written from the perspective of Lapulapu and the indigenous Filipino resistance to European colonialism, was erected in 1951 (Pinguel et al., 1998). The difference? In 1946 the Philippines gained independence and were no longer a colony of the United States.

Historical markers like these represent a sort of curriculum, one meant to communicate state-endorsed perspectives on the history and politics of a nation and the struggles over them (Loewen, 2000). In this case, the first plaque erected to honor Magellan was done so under the auspices of the United States' occupation of the Philippines, and, through taking on a Eurocentric view of Magellan's death, serves to reinforce the colonizer's justification for control of others' lands. This particular "curricular" outlook hence represents a form of what Huggins (1991) has referred to as a "master narrative" of history, one where indigenous and resistant voices and actions are distorted by those who have wrongfully taken power.

Conversely, the post-occupation plaque, which celebrates Lapulapu and Filipino resistance to European colonization, represents a curriculum of a different nature. It is not passive. It does not, as the first plaque does, let Magellan die of wounds, nor does it remain silent on Magellan's explicit purpose of coming to the Philippines. Rather, the second plaque honors a Filipino leader, and, as a state-sponsored historical marker of the newly independent Philippines, it also communicates a sense freedom.

The two plaques about Lapulapu and Magellan on the island of Mactan are a good example of the relationship between curriculum, power, and the politics of "official knowledge" (Apple, 2000), because each illustrates how, when knowledge is actively and purposefully constructed to convey meaning—as is true with all curricula—it is social, political, and, therefore, selective in purpose, content, scope, and perspective (Bernstein, 1996; Apple, 2004). By nature of its Eurocentrism and its selective use of perspective and language, the purpose of the first plaque was to tacitly support the colonization of the Philippines. The second plaque, meanwhile, was clearly constructed to communicate support for Philippine independence and resistance to foreign intrusion.

Indeed, like the plaques, all curricula seek to include and exclude, emphasize and de-emphasize, and embrace and isolate different content knowledge, different identities, and different politics. In this chapter, using examples from the United States and around the world, we seek to illustrate the inclusionary and exclusionary nature of curricula and how such curricular selectivity is directly linked to broader sociopolitical struggles. We begin this chapter with a brief discussion of what we mean by the term "curriculum," including a conceptual argument for the importance of curriculum within analyses of education. We then follow with some examples of how the rise in neoconservativism in countries such as the United States and Japan are manifest within struggles over the politics of inclusion and exclusion within the curriculum in both countries. Finally, we conclude this chapter with a discussion of how, even within the context of neoconservative curricular initiatives, akin to Lapulapu and those Filipinos who resisted colonization, curricular resistance exists. And it illustrates how educators interested in promoting equality and social justice operate with a different set of politics of curricular inclusion and exclusion.

The Curriculum

Before addressing the politics of curriculum, it is important for us to acknowledge what we mean when using the term "curriculum," and all that it implies. While there is no agreement within the field of curriculum studies on a single definition of "curriculum" (Beauchamp, 1982; Jackson, 1980; Kliebard, 1989), most of those in the field of education would at least agree that curriculum includes some set of content knowledge expected to be learned. However, the content of the curriculum also implicates pedagogy, because the selection of content is done with the express purpose of *transmitting* the said content (McEwan and Bull, 1991). The structure of knowledge is further implicated, since how we communicate ideas is also knitted together with the form in which those ideas are presented (Segall, 2004a, 2004b). Thus, to talk about curriculum means to talk about content, the form that content takes, and the pedagogy that guides the process of intended transmission (Apple, 1995; Au, 2007).

This more expansive view of the curriculum as embracing content, form, and pedagogy is important for the present analysis. Fundamentally, we assert that the curriculum is of central and immediate importance when we critically consider the politics of classroom knowledge because now we must be concerned not only with what is taught, but also how "what is taught" is structured and how "what is taught" is communicated. Put more simply, it is important to analyze the politics of the curriculum because to examine them is to simultaneously examine most aspects of classroom life.

The Politics of the Curriculum

One of the essential questions driving any curriculum is: What knowledge is of most worth? Teachers, departments, schools, districts, states, or federal authorities (or a combination of any and all of these bodies) then answer this question by outlining what knowledge will be considered legitimate or officially accepted within the classroom. The establishment of the boundaries of official knowledge is thus a political endeavor (see, e.g., Cornbleth and Waugh, 1995), one that is tightly knitted with social and political struggles happening outside of education (Apple, 2000, 2006; Bernstein, 1999). In this way the curriculum, and all it embodies, is a manifestation of the politics of inclusion and exclusion of not only content knowledge but of identities as well (Bernstein, 1996). Here we will focus on two examples of neoconservative curricular initiatives, one from the United States and one from Japan, to illustrate the ways in which the selection of curricular knowledge is driven by the politics of inclusion and exclusion.

The United States

Recent decades have witnessed the rise of the "conservative modernization" in the United States (Apple, 2006; Shor, 1986). Oftentimes lamenting a perceived decline in Western or "high" culture, neoconservatives, as one force within a larger rightist movement, have consistently advocated for what they believe to be a return to "traditional" knowledge, usually that knowledge historically associated with the Western European canon (Apple and Buras, 2006). The neoconservative impulse in the United States thus places high value on a specific curriculum, one that has a strong tendency to focus primarily on the cultures and peoples associated with northern and West European nation states and cultures to the exclusion of more multicultural, non-Western knowledge bases (Apple, 2006).

A key example of the neoconservative politics of exclusion and inclusion in the United States can be found within the Core Knowledge curriculum. The Core Knowledge curriculum grew out of E. D. Hirsch's (1987) widely read neoconservative text, *Cultural Literacy*, and represents many of the critiques of public education provided by Hirsch (1996) in *The Schools We Need and Why We Don't Have Them*. The Core Knowledge curriculum is a pre-k through eight grade curriculum that grows from the neoconservative assumptions of the "...hegemony of the left in major social institutions, the threat posed by countercultures to American Western civilization, and the fragility of national unity in light of identity politics..." (Buras, 2006, 46). The movement associated with the Core Knowledge curriculum possibly represents the most "successful" neoconservative educational reform initiative. As Buras (2008) reports, as of 2007 there are over 1,000 schools that officially participate in the Core Knowledge network. Roughly 500 of these are public schools, 300 are charter schools, and the remaining are private or religious schools.

A summary of the analysis of Core Knowledge texts provided by Buras (2006) serves to illustrate how the broader conservative modernization, particularly neoconservative, politics help shape what should be included or excluded in the curriculum. For instance, as Buras notes, in one sixth grade Core Knowledge history text on immigration in the United States,

> It is only European immigrants who speak—Asian immigrants are virtually absent—and when they do, they are only "allowed" to speak in ways that support the majority storyline . . . [of] an affirmative narrative of success and consensus in the United States. (66–67)

Even when the Core Knowledge curriculum attempts to include a more nonwhite, multicultural historical narrative, as in the case of a second grade textbook on African American leaders of the civil rights movement in the United States, Buras finds that, in a section that addresses racism and segregation,

> Unequal conditions prevail . . . but African Americans never encounter whites exercising power in direct or explicit ways. Violent southern segregationists did not work to help "all" Americans. And the struggle for integration saw much violence—namely that perpetrated by whites against blacks. Yet all of this is absent within a . . . frame that manages convey a story . . . of the black civil rights struggle, but without malevolent whites. (68–69)

Thus, we see how neoconservatives in the United States, such as those associated with Hirsch, have strived to develop a curriculum that is inclusive of a particular historical narrative and exclusive of such sticky issues as racism.

In the above examples, this narrative selectively gives voice to white immigrants to the United States—focusing on concepts of protestant work ethic, meritocracy, and assimilation, and also selectively positions whites in the Jim Crow south of the United States as unconnected, or passively connected at best, to the often violent maintenance of institutionalized racism. Both perspectives, as portrayed within the Core Knowledge texts, actively promote a unified America, and by extension a unified American culture, that is absent of the reality of identity politics and unequal social and economic relations that neoconservatives so regularly deny. The neoconservative Core Knowledge curriculum thus illustrates how the politics of curricular inclusion and exclusion are so tightly connected to broader social and cultural struggles, and that within the contexts of these struggles, education is seen as one realm within which to advance a neoconservative agenda. It also illustrates the creative ways in which dominant groups take up the language and expressed needs of oppressed groups and institute the safest possible reforms in response to them. In this way, hegemonic relations are not seriously challenged, and in this case the politics of race is made into something less pervasive and less structural (see Fraser, 1987; McCarthy, Crichlow, Dimitriadis, and Dolby, 2005; Gillborn, 2008).

Granted, the Core Knowledge curricular initiative does not exactly represent a *state driven* neoconservative reform. While such reforms do exist within education in the United States, amongst, for instance, struggles over what knowledge is included in state-adopted textbooks (Apple, 2000; Scott and Branch, 2006), Core Knowledge is more of a popularly neoconservative curricular reform that has been officially adopted by some state run public schools, as well as amongst a number of charter and private schools (Buras, 2008). This fact, however, does not negate the importance of Core Knowledge as an example of neoconservative curricular inclusion and exclusion. Rather, the Core Knowledge initiative must be seen as part of a broader, rightward, conservative political turn in the United States more generally (Apple, 2006; Apple and Buras, 2006; Buras, 2006), and this particular curricular thrust represents just one way neoconservatives seek to reform education in the United States.

The United States, however, is not alone in its rightward drift over the last few decades. Like a number of other nations, Japan too has seen a rise in neoconservativism, and this political shift has manifested in a series of educational reforms there that have sought to include specific historical knowledge and perspectives while excluding others. It is to Japan that we now turn to for a second example of how the politics of inclusion and exclusion are embodied by the curriculum.

Japan

In conjunction with significant neoliberal economic reforms in the late 1990s, Japan also experienced a radical shift in social and cultural politics. In the face of increasing economic instability and risk, neoconservatives in Japan seized upon popular feelings of instability and social anxiety to advance a social and cultural agenda of developing morality and patriotism amongst the Japanese population. One neoconservative curricular initiative generated in this context was a concerted effort to intensify the teaching of Japanese nationalism vis-à-vis state-regulated history textbooks (Takayama, 2007, in press; Takayama and Apple, 2008).

In April of 2005, the Japanese Ministry of Education sparked international protest when it approved a new edition of the middle school history textbook. At issue was the way the Japanese textbooks portrayed two atrocities committed by the Japanese Army during World War II: the Nanjing Massacre (also referred to as the Rape of Nanking) and the treatment of Korean and other Asian women taken as sex slaves (also referred to as "comfort women") for the invading Japanese army. The new textbook, developed by the neoconservative Japanese Society for History Textbook Reform, was written with the expressed intent of increasing Japanese nationalism by downplaying the atrocities committed by the Japanese Imperial Army during World War II (Takayama and Apple, 2008; Zhao and Hoge, 2006).

During the Nanjing Massacre, the Japanese army systematically looted the city of Nanjing, raped at least 20,000 to 80,000 Chinese women, and ultimately killed 300,000 Chinese civilians over the course of two months. Further, between 50,000 and 200,000 mostly Korean, but also Chinese, Filipino, Thai, Vietnamese, Malaysian, and Indonesian, women were taken by the Japanese army and forced into sexual slavery. The 2001 edition of the same text denied that the Nanjing Massacre ever took place and ignored the existence of the sex slaves, and the 2005 textbook referred instead to the Nanjing "incident" and de-emphasized the existence and experiences of the sex slaves (Zhao and Hoge, 2006).

In 2007, another similar controversy surrounding these issues emerged, as plans for further revisions to Japanese history textbooks included rearticulating the Japanese army's role in the 200,000 civilian deaths in the 1945 Battle of Okinawa. At the time of the battle, many Okinawans were given grenades and ordered by the Japanese army to commit suicide. Amongst the revisions recommended to the newest Japanese history text by the Japanese Ministry of education is that the role of the Japanese army in ordering these suicides be minimized (BBC News, 2007).

The neoconservative curricular initiative to downplay Japanese military wartime atrocities during World War II clearly illustrates once again how struggles over social and cultural issues manifest in the politics of inclusion and exclusion in the curriculum. Here we see how neoconservative state power translates into an educational reform agenda that seeks to uphold an historical perspective that paints a picture of a more innocent imperial Japan in order to promote a patriotic

Japanese nationalism (Takayama and Apple, 2008), a reform agenda that selectively denies the suffering of Chinese, Korean, Okinawan, and other victims of what many might otherwise consider to be criminal acts. Thus, the neoconservative curricular reforms in Japan selectively include certain historical perspectives, ones that support an image of a more righteous Japanese state while simultaneously excluding the historical perspectives of those that were killed or oppressed through the brutality of a nationalist Japanese state and its military apparatus.

CURRICULAR RESISTANCE AND APPROPRIATION

It is important to recognize that, despite the rightward political turn and the rise of neoconservativism in both the United States and Japan, the politics of inclusion and exclusion embedded within these neoconservative curricular initiatives have been both appropriated and resisted. In the case of the Japanese textbooks, in 2005, Japanese and other scholars criticized the neoconservative curricular revisions regarding the Nanjing Massacre and the sex slaves (Zhao and Hoge, 2006), and in 2007, over 110,000 Okinawans protested the proposed textbook changes regarding the civilian "suicides" during the Battle of Okinawa (Associated Press, 2007).

Further, in the case of the Core Knowledge curriculum, despite its narrow structuring of American culture, some Core Knowledge teachers have used the curriculum in ways that may seem oppositional to the neoconservative politics embedded within it. For instance, one middle school teacher in the state of West Virginia made use of the Core Knowledge guidelines for teaching about the civil rights to attempt to interrupt the influence that she felt a nearby white supremacist, neo-Nazi compound might have on her students' understanding of race relations. In another instance, an indigenous American school on the Pine Ridge Indian Reservation in South Dakota uses the Core Knowledge curriculum alongside a Lakota studies and language program. And in yet another instance, the Core Knowledge curriculum is being used as part of a two-way bilingual/bicultural education program in Florida (Buras, 2008).

The above examples do not counter the arguments we have made here. Rather, they highlight the fact that historically any curriculum is socially and politically contested terrain (see, e.g., Kliebard, 2004), and that what finally manifests itself can sometimes either be a curricular compromise to appease critics (Apple, 1988; Apple and Buras, 2006) or a recontextualized curriculum that teachers take and adapt to their own classroom practices (Apple and Beane, 2007). Part of the power of the curricular initiatives discussed here, however, is that neoconservatives have been in a position to help establish what perspectives and content get included as common sense and acceptable within pedagogic discourse (Apple, 2006; Bernstein, 1996), thus further constraining the possibility of critics or classroom practitioners to have their inputs included into the curriculum.

A Progressive Politics of Inclusion and Exclusion

While the neoconservative politics of inclusion and exclusion tend to embrace nationalism, patriotism, and common culture while eschewing issues of multiculturalism, difference, and structural inequalities, there are other curricular initiatives that embrace an entirely different set of politics. These initiatives seek out to actively recognize social, cultural, and economic inequalities and work to address them by fostering conscious collective action amongst students and communities.

The United States

In the United States, there are several available examples of curricular initiatives developed around a politics of inclusion and exclusion that seek to foster social justice. For instance, in Milwaukee, Wisconsin, a public elementary school named La Escuela Fratney was started by a group of parents, teachers, and community activists. Through a grassroots, sustained community-based organizing campaign, this group pushed the Milwaukee Public School district to support the opening of this unique elementary school (Peterson, 2007).

Located in one of Milwaukee's most diverse neighborhood, La Escuela Fratney is a school whose curricular vision is based on the following program: a two-way, bilingual program in Spanish and English; a multicultural, antiracist curriculum; whole language and natural second language learning; cooperative learning and discipline; a thematic curriculum approach; critical thinking; local governance of the school by teachers and parents; significant parent involvement; and substantial links to the community (Peterson, 2007). Despite struggles to maintain their radically progressive vision of public education, and in contrast to the politics of neoconservativism, La Escuela Fratney manages to establish a politics of curricular inclusion that embraces community empowerment, antiracism, and authentic multiculturalism.

Another example comes from the Cabrini Green low-income housing projects in Chicago. Cabrini Green is notorious both for high incidences of crime, including robberies, drug dealing, and gang activity, as well as for being so poorly maintained by the Chicago Housing

Authority that many of its residences were declared unlivable (Schultz, 2007). In short Cabrini Green stands as a monument to economic neglect and the lack of adequate social programs in the United States.

As a fifth grade teacher in Chicago, Brian Schultz explains, he uses the inadequate socioeconomic infrastructure in and around Cabrini Green as the basis for his curriculum. Students in his class identified a litany of problems in their community that needed to be addressed, among them, the poor quality of their school facilities. Schultz and the students in his class collaborated to develop a plan of action to try and get the physical condition of their school improved. This plan, which became the center of their curriculum for the semester, included drafting letters and e-mails and sending them out en masse to public officials, community members, and the media (Schultz, 2007).

Through this curriculum of social justice and social action, as public response to their project grew, Schultz's students developed a stronger sense of empowerment and produced work of such a quality some officials disbelieved it to be theirs. As Schultz (2007) himself notes:

> The project was inspirational to all who came in contact with the young activists. The best part to me was that the kids...were able to engage in such meaningful work, creating a counternarrative, since many people simply could not believe that the students responsible were "inner-city kids" that were doing "such amazing work"...(77)

Schultz's comment is important here. Students in his classroom produced a "counternarrative" in their curricular initiative because they, in taking part in a project that they felt was important and powerful, did not embody the stereotypes many hold of urban youth as disaffected, unsuccessful learners. In this sense, their curricular initiative embraced a politics of inclusion and exclusion that not only sought equitable and just social and educational transformation, but also sought individual and personal transformation—reflecting a politics of democratic, progressive activism in the process (see also Gutstein, 2006).

In both of these examples from the United States, we see no romantic image of "America," no calls for a muted or assimilationist multiculturalism, and no masking of inequalities. Rather, at a school such as La Escuela Fratney, we are offered an example of community empowerment and social justice education, and in Schultz's fifth grade class, children experience a curriculum that critically acknowledges inadequate funding of schools while encouraging student activism to address the same inadequacy.

South Korea

As another example, teachers in South Korea, as part of their successful drive to form and gain legal recognition for the Korean Teachers Union (KTU) in the face of an increasingly conservative government, implemented a series of lesson plans called *Gongdong* cooperative lessons. These lessons represented an alternative that unionized teachers taught in their classrooms forty times since 1997 in a coordinated effort to advance students' understanding of issues critical to social justice in South Korea. Topics of *Gongdong* lessons have included labor law, the rights of people with disabilities, antiwar, anticorruption of public officials, human rights, Korean reunification, and critical issues associated with the presence of U.S military forces based within South Korea (Kang, in press).

The *Gongdong* lessons are a direct extension of the labor organizing undertaken by the KTU and operate in direct opposition to the politics of the increasingly conservative South Korean government. As such, *Gongdong* lessons embrace a critical theoretical approach to education that actively seeks to challenge the politics of inequality within South Korean society while encouraging and fostering action toward social justice on the part of both students and teachers (Kang, in press).

Similar to the examples from the United States, the South Korean case represents a particular set of politics of curricular inclusion and exclusion. In this case, in an organized act of resistance to government policies that they felt were detrimental to teachers, students, and society, the teachers of the KTU implemented a curriculum that was inclusive of the politics of social justice, equality, and activism, and one that was at least partially exclusive of government-endorsed concepts. Despite the fact that the *Gongdong* lessons were temporary and finite in that they did not represent a reform of the curriculum of the entire educational system in South Korea, they still represent an important attempt on the part of teachers to challenge curricular hegemony in their country.

Namibia

Yet another example comes from the African country of Namibia, where, after the South West Africa Peoples Organization established the country's first democratic government in 1990, leaders attempted to reform their educational system to challenge the racist apartheid philosophy that guided it during colonial rule. As such, the new Namibian government publicly committed itself to educational programs, a new curriculum, if you will, that promoted equal access to an educational system that strove to be fair and equitable for all students, and an educational system that placed importance on student-centered learning and democratic decision-making (Dahlström, Swarts, and Zeichner, 1999).

The Namibian government attempted to make these educational reforms through the restructuring of teacher education, with the intent of training new teachers to build the capacity to make such reforms successful. Thus, as an extension of an overall social justice agenda, a method of teacher education called "Critical Practitioner Inquiry" (CPI) was developed and implemented with Namibian teachers. CPI is an approach to teacher education that is based in critical pedagogy, one that asks teachers to critically interrogate their own practices to develop an understanding of the ways that social inequalities become embedded in education. Further, it falls into a tradition of teacher critical inquiry "that seeks to address the production and reproduction of inequalities in society and to promote learner-centered and democratic classroom practices" (Dahlström et al., 1999, 156).

As with the South Korean case, here the politics of curricular inclusion and exclusion are important. In a postcolonial, postapartheid-like context, a curriculum of Namibian liberation and democracy took shape. In this case, the Namibians embraced a politics of social justice and equality, a politics that saw education as one means of correcting the inequalities established and imposed upon them by their oppressive colonizers. Again, we also see how the Namibians, in direct response to the inequality of the colonial system of education (and their exclusion from decision-making power within that system), developed a curriculum that was inclusive of core concepts of equality, equity, and democratic transformation.

Conclusion: The Curriculum as a Continuing Site of Struggle

In this chapter we have sought to make a simple argument: The curriculum is a manifestation of the politics of inclusion and exclusion, and that these politics represent broader social and political struggles taking place outside of schools. Taking for granted that we could not include examples from everywhere around the world in our analysis, we chose a few select examples to illustrate how this process takes place. In the case of the United States, we highlighted the neoconservative Core Knowledge curricular initiative, one that, despite paltry efforts to include the voices of nonwhite populations, ultimately weaves together a narrative of a country united by a common culture, one rooted in patriotic notions of meritocracy and the classical Western traditions, as well as a country devoid of the realities of structural inequalities such as racism, classism, sexism, or homophobia. We also drew on a similar rightward turn in Japan, where neoconservatives have sought to reform Japanese textbooks to downplay the atrocities committed by the Japanese military during World War II. In both cases, anything that threatened to rupture the romantic image associated with neoconservative traditional values was either summarily left out or sculpted to fit within the neoconservative historical narrative.

In highlighting these more conservative curricular initiatives, however, we also sought to make another important point: The curriculum is always contested or reappropriated at some level. The Japanese Ministry of Education has faced outright protest over its proposed textbook changes, and the Core Knowledge curricula have often been adapted and used by teachers and schools for their own means—means that sometimes might even contradict the thrust of the Core Knowledge initiative.

Such contested and reappropriated curricula also direct us to another crucial point: The curriculum is a site of struggle. As such, we offered examples of curricular initiatives that embraced a different set of politics from that of the neoconservatives. The examples from the United States demonstrated how school-wide curricular reforms were built on a politics of social justice—and where students, teachers, and schools were active in the process of education. In these instances, and so many more to which we could point, the goal was for students to develop a consciousness that might be considered counter-hegemonic to that fostered by the neoconservatives. The South Korean example further demonstrated how a broader movement for labor rights amongst teachers developed into the manifestation and implementation of resistant *Gongdong* lessons, which became a space carved out of the state-mandated curriculum where teachers could pursue a politics of equality and democracy. Finally we highlighted how in Namibia a newly formed popular democratic government sought to develop a curriculum that fostered access to equitable education as well as democratic participation vis-à-vis preservice and in-service teacher education.

In all of these cases there is a clear sense of potential for the possible power of educators to develop a politics of curricular inclusion and exclusion that seeks positive social change, and which recognizes

oppression and structural inequalities while promoting critical thinking aimed toward social transformation and recognizing the crucial
role of the curriculum in the development of critical consciousness
amongst learners. We should not romanticize this situation, however.
Curricular struggles, like so many other struggles over knowledge
and power, are not easy, and there is no guarantee of victory in building an education worthy of its name. But the examples of counter-
hegemonic curricula we have given here do demonstrate that those
in dominance who say that "There is no alternative" ("TINA") are
wrong. There definitely are alternatives—and they are being built by
real teachers and real schools throughout the world.

In a sense, all of the curricular initiatives we've discussed in this
chapter raise the same questions: Whose knowledge is of most worth?
How is this sense of worth connected to social, cultural, and political
struggles? What role can and does the curriculum play in educational
and social transformation? Or, put differently, do the politics of our
curriculum side with the romantic past of Magellan the "explorer"?
Or, do the politics of our curriculum side with the struggle of
Lapulapu against Magellan the invading colonizer?

References

Apple, M. W. (1988). "Social Crisis and Curriculum Accords." *Educational
Theory*, 38(2), 191–201.
———. (1995). *Education and Power* (2nd ed.). New York: Routledge.
———. (2000). *Official Knowledge: Democratic Education in a Conservative
Age* (2nd ed.). New York: Routledge.
———. (2004). *Ideology and Curriculum* (3rd ed.). New York: Routledge.
———. (2006). *Educating the "Right" Way: Markets, Standards, God, and
Inequality* (2nd ed.). New York: Routledge.
Apple, M. W. and Beane, J. A. (eds.) (2007). *Democratic Schools: Lessons in
Powerful Education* (2nd ed.). Portsmouth, NH: Heinemann.
Apple, M. W. and Buras, K. L. (2006). "Introduction." In M. W. Apple and
K. L. Buras (eds.), *The Subaltern Speak: Curriculum, Power, and Educa-
tional Struggles*. New York: Routledge. 1–39.
Associated Press. (2007). "Okinawa Residents Protest Over WWII History
Textbook Amendment." *The International Herald Tribune*. Retrieved
October 4, 2007, from http://www.iht.com/articles/ap/2007/09/29/
asia/as-gen-japan-okinawa-demonstrations.php.
Au, W. (2007). "High-stakes Testing and Curricular Control: A Qualitative
Metasynthesis." *Educational Researcher*, 36(5), 258–267.
BBC News. (2007). "Okinawa Anger at Textbook Plans." *BBC News*.
Retrieved October 3, 2007, from http://news.bbc.co.uk/go/pr/fr/-/2/
hi/asia-pacific/6229256.stm.

Beauchamp, G. A. (1982). "Curriculum Theory: Meaning, Development, and Use." *Theory Into Practice*, 21(1), 23–27.

Bernstein, B. B. (1996). *Pedagogy, Symbolic Control, and Identity: Theory, Research, Critique.* London: Taylor and Francis.

———. (1999). "Official Knowledge and Pedagogic Identities." In F. Christie (ed.), *Pedagogy and the Shaping of Consciousness: Linguistic and Social Processes.* New York: Cassell. 246–261.

Buras, K. L. (2006). "Tracing the Core Knowledge Movement: History Lessons from Above and Below." In M. W. Apple and K. L. Buras (eds.), *The Subaltern Speak: Curriculum, Power, and Educational Struggles.* New York: Routledge. 43–74.

———. (2008). *Rightist Multiculturalism: Core Lessons on Neoconservative School Reform.* New York: Routledge.

Cornbleth, C. and Waugh, D. (1995). *The Great Speckled Bird* (1st ed.). New York: St. Martin's Press.

Dahlström, L., Swarts, P., and Zeichner, K. (1999). "Reconstructive Education and the Road to Social Justice: The Case of Post-Colonial Teacher Education in Namibia." *International Journal of Leadership in Education*, 2(3), 149–164.

Fraser, N. (1987). *Unruly Practices.* Minneapolis: University of Minnesota Press.

Gillborn, D. (2008). *Racism and Education: Coincidence or Conspiracy?* London and New York: Routledge.

Gutstein, E. (2006). *Reading and Writing the World with Mathematics.* New York: Routledge.

Hirsch, E. D., Jr. (1987). *Cultural Literacy: What Every American Needs to Know.* New York: Vintage Books.

———. (1996). *The Schools We Need and Why We Don't Have Them.* New York: Doubleday.

Huggins, N. I. (1991). "The Deforming Mirror of Truth: Slavery and the Master Narrative of American History." *Radical History Review*, 49, 25–48.

Jackson, P. W. (1980). "Curriculum and Its Discontents." *Curriculum Inquiry*, 10(2), 159–172.

Kang, H. R. (in press). In M. W. Apple, W. W. Au, and L. A. Gandin (eds.), *The Routledge International Companion to Critical Education.* New York: Routledge.

Kliebard, H. M. (1989). "Problems of Definition in Curriculum." *Journal of Curriculum and Supervision*, 5(1), 1–5.

———. (2004). *The Struggle for the American Curriculum, 1893–1958* (3rd ed.). New York: Routledge Falmer.

Loewen, J. W. (2000). *Lies across America: What Our Historic Sites Get Wrong.* New York: Touchstone.

McCarthy, C., Crichlow, W., Dimitriadis, G., and Dolby, N. (2005). *Race, Identity, and Representation in Education* (2nd ed.). New York: Routledge.

McEwan, H. and Bull, B. (1991). "The Pedagogic Nature of Subject Matter Knowledge." *American Educational Research Journal*, 28(2), 316–334.

Peterson, B. (2007). "La Escuela Fratney: A Journey Towards Democracy." In M. W. Apple and J. A. Beane (eds.), *Democratic Schools: Lessons in Powerful Education*. Portsmouth, New Hampshire: Heinemman. 30–61.

Pinguel, B., Wei, D., and Shalom, S. R. (1998). "Reframing the Spanish-American War in the History Curriculum: History and Point of View." In D. Wei and R. Kamel (eds.), *Resistance in Paradise: Rethinking 100 Years of U.S. Involvement in the Caribbean and the Pacific*. Philadelphia: American Friends Service Committee. 1–24.

Schultz, B. D. (2007). "Feelin' What They feelin': Democracy and Curriculum in Cabrini Green." In M. W. Apple and J. A. Beane (eds.), *Democratic Schools: Lessons in Powerful Education*. Portsmouth, New Hampshire: Heinemman. 62–82.

Scott, E. C. and Branch, G. (eds.) (2006). *Not in Our Classrooms: Why Intelligent Design Is Wrong for Our Schools*. Boston, MA: Beacon Press.

Segall, A. (2004a). "Blurring the Lines between Content and Pedagogy." *Social Education*, 68(7), 479–482.

———. (2004b). "Revisiting Pedagogical Content Knowledge: The Pedagogy of Content/the Content of Pedagogy." *Teaching and Teacher Education*, 20, 489–504.

Shor, I. (1986). *Culture Wars: School and Society in the Conservative Restoration 1969–1984*. Boston: Routledge and Kegan Paul.

Takayama, K. (2007). "A Nation at Risk Crosses the Pacific: Transnational Borrowing of the U.S. Crisis Discourse in the Debate on Education Reform in Japan." *Comparative Education Review*, 51(4).

———. (2008). "Japan's Ministry of Education 'Becoming the Right': Neoliberal Restructuring and the Ministry's Struggle for Legitimacy." *Globalisation, Societies and Education*, 6(2), 131–146. Philadelphia, PA: Routledge.

Takayama, K. and Apple, M. W. (2008). "The Cultural Politics of Borrowing: Japan, Britain, and the Narrative of Educational Crisis." *British Journal of Sociology of Education*, 29(3).

Zhao, Y. and Hoge, J. D. (2006). "Countering Textbook Distortion: War Atrocities in Asia, 1937–1945." *Social Education*, 70(7), 424–430.

CHAPTER 7

Starting with What the Children and Their Families Know: An Argentine Elementary School Addresses Issues of Underperformance of Bolivian Immigrant Children

Mary Q. Foote and Mónica Mazzolo

INTRODUCTION

The significant numbers of immigrant children in schools in the United States warrant a focus on their specific educational needs. Because many of these children are both poor and English learners, their educational needs intersect with other marginalized groups in the United States, such as native born Latinos, Native Americans, and African Americans. What we can learn from examining the strengths and needs of immigrant children can support schools in building on those strengths and addressing their needs; it can as well support schools in building on the strengths and addressing the needs of marginalized groups of students more generally, and supporting more inclusive school practices. This chapter examines a project initiated in an elementary school in Luján, Argentina, to address the under-achievement of Bolivian immigrant children when compared with their Argentine peers. Thus, this chapter provides details about the project and considers the following outcomes of the project: (a) the participation and achievement of the Bolivian children; (b) parent participation in the school; and (c) teacher learning.

BACKGROUND

One of the authors of this chapter, Mónica Mazzolo, is the director of Escuela 10 (Public School 10), the school being examined in this chapter. It was she who supported the faculty of Escuela 10 when they expressed a desire to confront the issue of the underachievement of the Bolivian immigrant children in their school. The other author, Mary Foote, lived in Argentina and visited Escuela 10 on multiple occasions during the first six months of the school year during which the project was undertaken. The two authors met in Argentina in 2006 to revisit the project that had been implemented in 2001.[1] Using a set of semi-structured interview questions, Foote explored with Mazzolo the impetus for the project, its development, and its results. The conversations, which were in Spanish, were tape-recorded. Foote later took notes in English during reviews of the tape-recorded material. These notes, as well as e-mails between the two authors, were used as the basis for describing and contextualizing the project for this chapter.

LOCAL CONTEXT

Luján, Argentina, a city of approximately 100,000 inhabitants, is located in the Province of Buenos Aires about an hour west of the city of Buenos Aires. There are thirty-three public elementary schools in Lujan serving about 20,000 students. In the Province of Buenos Aires, students in grades one through six attend the elementary schools. Upon graduation from the elementary (Basic Primary) school, students attend three years of middle school (Basic Secondary), which, along with elementary schooling, is obligatory. This course of study is followed by three years of high school.

Primary schools in Argentina are structured in two ways. There are schools that children attend for the entire day, and there are schools where there are two sessions per day, one in the morning from about 8:00 A.M. to noon and one in the afternoon from about 1:00 P.M. until 5:00 P.M. Under the latter system, which is the more traditional in Argentina, children attend either the morning or the afternoon session. Teachers may teach one or both of the sessions, although the director (principal) is in charge of both. The planning and organization of the school, therefore, is managed on a session basis, with faculty meetings being held separately for the morning and afternoon faculties. Escuela 10, the school whose project is being documented in this chapter, is a school that operates on the traditional Argentine model with both morning and afternoon sessions. The school enrolls approximately 360 students. There is one class at each grade level, grade one through grade six, during both the morning and afternoon sessions.

Escuela 10 is located on the outskirts of Luján in a semirural area where there is a mix of primary residences and weekend homes. The first Bolivians to come to the area came as seasonal migrant farm workers to work on the grounds of these weekend homes. Growing vegetables and fruits, if one has the land, is an important aspect of Spanish culture, one that has been adopted in the Western Hemisphere as well, and many of these weekend homes have their own gardens, sometimes substantial ones, providing seasonal employment opportunities for gardeners. Escuela 10 has enrolled the children of these Bolivian migrant workers since before the current director, Mazzolo, came to the school. More recently these migrant workers have begun to settle permanently in the area. In addition, beginning six or seven years ago, the population of Bolivian immigrants in Argentina in general and in this area in particular increased due to problems with the Bolivian economy. This led to an increase in the percentage of Bolivian immigrant students attending Escuela 10. In addition, this group of Bolivians who came to the area settled down and established a permanent presence in the area and the school. No longer did the majority of Bolivian children attend school only during the growing and harvest seasons and leave thereafter, possibly to return the following year for a period of time. Currently, Bolivian immigrant children make up approximately 15 percent of the school population.

Educational Context

There were many aspects in the way the school operated in the years that Mazzolo was the director that may have positioned the school to be ready for and alert to the needs of the Bolivian immigrant children who had so recently become a consistent part of the student body. As a classroom teacher in the days before she became a school director, Mazzolo had incorporated investigations of indigenous people with the study of the Spanish involvement in Argentina and, more broadly, Latin America. She conceptualized her Social Studies program as an encounter between two cultures. The arrival of the Spanish in Argentina was viewed neither as a triumph nor a tragedy. Since without the arrival of European peoples the majority of the teachers would not be there, the "discovery" of America was not seen as tragic. But neither was it seen as triumphant, since the indigenous communities who lived in the lands prior to the European arrival had a cultural richness of their own. This richness, which in South America as well as in North American has often been ignored or diminished, was something that was taken up and valorized by Mazzolo in her own classroom teaching. Although this interest in foregrounding indigenous culture may be somewhat unusual in urban settings in Argentina, it is not unusual that a high value is placed on indigenous culture in schools in the interior of the country, where the presence of indigenous population is more.

When she arrived at Escuela 10 to assume the directorship in 1996, Mazzolo found a group of teachers who also thought that the validation of the indigenous experience in Argentina was important. At the same time, Mazzolo supported the development of a culture of communication and shared knowledge among the teachers in the school. It was a priority for teachers to know what was being taught at the other grade levels. Faculty meetings at the start and conclusion of each school year planned for and then evaluated the effectiveness of the curriculum that had been implemented that year. For example, it was common in the planning meeting for teachers to brainstorm about how to use events that were happening or would be taking place during the up-coming school year to motivate the children. The school operated from a common belief that children learn better from situations that interest them than from situations that don't.

The teachers in this school have therefore historically looked at issues that are occurring in the world outside of the community and the nation, which would interest the children. They address the required content through a study of these issues. They realize that they may make some mistakes in addressing content in this manner, but they believe as well that if the child isn't interested, he or she won't learn. This focus on the richness inherent in the indigenous communities of Argentina, as well as the knowledge base of the children's curricular trajectory as they move through the school, and their method of addressing content through issues that interest particular children in their school may have positioned these teachers to include a focus on Bolivia and Bolivian culture in their curriculum.

The Project of Cultural Integration

The project to address the underachievement of the Bolivian children with respect to their Argentine peers did not begin with a clear plan. Instead it began with a concern voiced at first by a single teacher. At the faculty meeting to discuss curriculum plans for the 2001 school year, one teacher in particular voiced a concern that the Bolivian children in the school were, generally speaking, very quiet, that they didn't participate to the same extent as the Argentine children, that their parents' weren't as involved, and that they weren't performing academically at levels comparable to the Argentine children. Some of the other teachers seconded this concern. The teachers felt that it was their fault that they weren't reaching the Bolivian children. The locus of responsibility for change was therefore centered in the school and among the teachers. This is not a typical point of view held by teachers in Argentine schools. As in the United States, informal talk among teachers in Argentina often positions the families as responsible for school failure (Foote, 2001). The teachers in this school thus began their project of cultural integration from a position of believing themselves to be the ones who had to make changes. They began by looking at what they could actively do to change the curriculum to address the needs of the children.

At the planning meeting for the 2001 school year, teachers identified two areas of concern about the Bolivian children: the language gap and the differences in culture. The Bolivians often spoke indigenous languages at home; the mothers particularly were often more comfortable with these native languages than with Spanish. Because of this language gap, communication between the school and the Bolivian parents was at times difficult. It also appeared that both the children and their parents felt somewhat adrift in a foreign land. They were becoming less connected with their own heritage and at the same time felt alienated from Argentine culture. From the conversations at this planning meeting, the project of integration was born. The teachers decided to launch a project to integrate the Bolivian immigrant children more fully into the school, and they put their plans in writing. Following the communal manner in which the school is directed, this written plan was circulated among the faculty for further comments, and when everyone had signed on, the teachers began work on the project. Although there were some teachers who were much more involved than others, everyone agreed to work on the project and all contributed ideas. Some teachers were more passionate about the project, and they carried along the others with their enthusiasm.

The project developed over the course of the academic year of 2001. It was envisioned as a project of cultural integration: an integration of elements of both Argentine and Bolivian history and culture into the school curriculum. Curriculum was developed to explore the literature, music, traditional clothing and food, as well as the history and geography of each country. One of the initial steps taken in the implementation process was to hold a parent meeting to communicate with both Bolivian and Argentine parents about the project. From the onset, both the Argentine and Bolivian parents were involved and both contributed suggestions for the project.

The project built on work that was traditionally done in the school. For example, the school routinely examined the history and culture of indigenous Argentine peoples as a part of the curriculum, supporting the understanding that it is the indigenous people who are the true keepers of this land, which belonged to them before colonization. As part of the project, the examination of Bolivian peoples and indigenous culture was added to this extant examination of indigenous Argentines. In the lower grades, stories and legends, myths and music from both countries were examined. Vocabulary in the multiple indigenous languages of the children was shared and validated. Parents were enlisted to speak to the classrooms about festivals and ceremonies that are typical of Bolivia, such as the Carnival de Oruro. Teachers made contact with parents when they dropped their children off at school, using that time to establish a date when a parent could come and talk with the class. The Bolivian parents eagerly responded to the invitation to come into the school to share their knowledge. As Mazzolo notes, many books that are written about indigenous ceremonies are written by nonindigenous people, so the mothers coming to share their insider knowledge lent authenticity to the study of these festivals. And the Bolivian children, some of whom had rarely spoken in school, responded to the parents' presence and presentation of information about festivals and ceremonies, with which they too had firsthand experience, by talking extensively. It turned out they had a lot to say.

In the upper grades, where the curriculum took a more historical focus, teachers began to look at issues such as a comparison of the declarations of independence of Bolivia and Argentina, noting similarities and differences. When the students were studying Argentina's fight for independence from Spain, they also examined the Bolivian fight for independence. Colonization and the fight for independence in Argentina were compared to those of Bolivia. For example, the children learned that under Spanish colonial rule both countries were a part of the important Spanish Viceroyalty of Rio de la Plata, administered from the city of Buenos Aires.

From the start, there was an attempt to integrate all curricular areas into the project. Art and music and literature formed the basis of the integration in the early grades. These were continued in the upper grades where other opportunities for integration were also explored. For example, when the students were studying maps of Argentina and Bolivia, they would study percentages in mathematics and discuss percentages of the populations of the two countries that belonged to various indigenous groups. The study of proportion was linked to the study of scale in mapping as well.

The project thus developed throughout the course of the academic year, relating Bolivian themes to the Argentine ones that had traditionally been a curricular focus. The school year traditionally ends in Argentine primary schools with an end-of-year performance. This is an important event in Argentine school life. The performance at the end of the 2001 school year focused on typical music and dancing of both Bolivia and Argentina. The Physical Education and Art teachers contributed their expertise to the scenery, costumes, and dance. The parents were also asked to contribute their expertise and both Bolivian and Argentine parents were very involved.

FRUITS OF THE PROJECT OF CULTURAL INTEGRATION

The impact of this project can be seen in three areas: (a) increased achievement and participation of the Bolivian children, (b) increased participation of their parents in the life of the school, and (c) significant learning by teachers.

Achievement and Participation

Although the data to support the claim that the project was a success in increasing the participation and achievement of the Bolivian immigrant children is largely anecdotal, it is nonetheless compelling. The incorporation of their experiences into the school curriculum and the validation of their experiences as worthy of attention in school seem to have encouraged the students to add their voice to the classroom conversation. The teachers report that both the Bolivian and Argentine children now perform at the same level academically. That is to say, they can no longer predict by their nationality which children will perform at particular levels. This is just what Rochelle Gutiérrez (2002) suggests would be a measure of whether schools had achieved equity.

The teachers also report that the classroom participation of the Bolivian immigrant children is now comparable to that of their Argentine peers. Formerly 'silent' children began to participate with enthusiasm when the content was something that was meaningful and familiar to them. They were encouraged to add their voices to the classroom conversation when members of their own community, people they recognized as like themselves, were invited into the school. When their knowledge was honored and validated, they began to recognize themselves within the curriculum. The teachers joke that they managed the integration of the Bolivian immigrant children badly because they now talk too much. The greater participation and improved performance of the Bolivian immigrant children is reminiscent of the findings that the funds-of-knowledge research group has reported. Student learning is enhanced when teachers become informed about the homes and communities of children who are unlike them and bring that knowledge into the school setting (González, Andrade, Civil, and Moll, 2001; Moll, Amanti, Neff, and Gonzalez, 1992; Moll and Gonzalez, 2004).

Parent Involvement

Parents too came to view themselves as part of the school community in ways that they previously did not. This again supports findings of the funds-of-knowledge group (Civil, 1998, 2000; Civil and Bernier, 2004). One particularly poignant case is that of a mother whose child had broken his leg in Bolivia, had been poorly attended to medically, and walked with a severe limp. The mother was initially very reticent to speak to the teachers, rarely talking to them except to answer "yes" or "no" to questions. The year after the project began, she began to talk more to her son's teacher and the teacher began to encourage her to seek medical attention for the child. The mother took these ideas and suggestions home to her husband, and he began to take the child to various doctors. At a children's hospital, the doctors said that with an operation they could correct the problem. The father came to school at this point to ask the director whether she thought this was a good idea. Mazzolo's interpretation of this request by the father is that because he felt supported by the school, he could come to discuss the matter with people who he felt had his son's best interests in mind. The parents decided to go ahead with the operation, and during the three months that the child was out of school the parents regularly came to the school library to get him books to read and to report on his progress. Mazzolo identifies this as one of the fruits of the project. And this case does not stand alone. Other parents now demonstrate as well that they see the school as a resource. Generally speaking, the Bolivian families are more integrated into the school community. Before the project began, few Bolivian parents attended Parent Organization meetings. Now they attend these meetings in significant numbers. They also participate in classroom activities at times other than when their expertise is needed.

Teacher Learning

Teacher learning has also been a result of the project. One aspect of this learning is a deeper appreciation of the need to examine who their students are and what knowledge they bring to school. Through this project, the teachers learned how to expand the curriculum development skills they already employed to include a routine examination and integration of issues related to Bolivia. Now instead of focusing only on Argentine history, geography, and culture when exploring integrated topics of study, they routinely consider how to integrate Bolivian history, geography, and culture as well. In a study in which she examined the results of forging links between teachers and children from a different cultural background, Foote (2006) also found that such efforts could support teacher learning, which then resulted in teachers changing their set practices.

THE PROJECT CONTINUES

The school continues to address curricular goals through projects that interest the children. One thing they have focused on this year is the World Soccer Cup. They predicted that the children, along with the entire country, would be consumed by this event, and that it would be an opportunity to connect with material that the children would find very engaging. The teachers have used the Mundial, as the World Cup is known in Spanish, to explore and compare the history, geography, and culture of the various countries whose teams participated. They looked at such things as folk tales and legends, as well as the land area of the various countries, and the comparative sizes of their populations, to name only a few topics. Now that the Mundial has finished, they are devoting time during the balance of this year to engaging in a project exploring the risks and benefits of large pulp mills that the paper industry is thinking of building on the Paraná River. This has become a contentious topic throughout the nation because of its potential for water pollution.

And so, their tradition of teaching through projects that engage and interest the children continues. It has become routine for teachers at the beginning of the year meetings to consider more carefully who the students are. They are more focused on incorporating and highlighting issues that are of importance to the Bolivian community as the number of Bolivian children in the school continues to be significant. The recent election of an indigenous Bolivian as president of the country is a case in point. The Bolivian children were particularly interested in the new president of their homeland because he comes from a background like their own. The children wanted to write him letters. The teachers mobilized to support this communication. They contacted the Bolivian Embassy in Buenos Aires in order to investigate a way to direct the children's letters to someone who would respond to them.

DISCUSSION

Some might look at the accomplishments of the Project of Cultural Integration and consider it complete. During the course of the 2001 school year, the teachers explicitly addressed and integrated Bolivian cultural and historical information into what had previously been a predominantly Argentine focused curriculum. Bolivian parents were enlisted and their knowledge validated to such a degree that they became visibly integrated into the school community. Achievement and participation by both Argentine and Bolivian children is now equal. The success of the project stands as a testament to the positive effects of an effort directed at more inclusive schooling. Yes, these teachers and families accomplished many things, but the work of the project continues to date, and this work must be continuous if schooling is to remain inclusive of all students. Cultural and historical issues important to the Bolivian community continue to be considered and incorporated into whatever topics are being used as the school's core curriculum. The Bolivian perspective is one that the teachers routinely consider when developing their curriculum plans. Instead of the project being a one shot attempt to address discrepancies in participation and performance, it has become a permanent way of working for the school faculty. Mazzolo notes that if a group of Peruvian children showed up tomorrow, the teachers would begin exploring how to incorporate aspects of Peruvian history and culture into their curriculum. The way the teachers now plan has evolved to address issues of inclusivity for all students head-on. The success of the project can surely be attributed to the successful inclusion of material about Bolivia into a curriculum that had historically paid attention primarily to Argentina. The success of the project may also be due to structures that were already in place at the school and that supported the inclusion of content that was significant to the Bolivian immigrant children.

One reason for the project's success was surely the careful attention that the teachers paid to cultivating relationships with Bolivian parents and using them as academic resources in the school. Without this valorization and use of the information base that the Bolivian parents possessed, it is hard to imagine that the Bolivian community would have been so successfully engaged to support their children's academic success. At the same time, it seems as if the general methodology that the teachers had historically relied on for curriculum development provided a platform from which they could take on this new project of attending to issues of inclusive schooling. The teachers were already accustomed to considering and valuing the indigenous Argentine experience, so that extending that to include indigenous Bolivian peoples was not akin to venturing into uncharted territory. The same can also be said with regard to the integration of curricular areas. Integrating the Bolivian cultural and historical pieces was not a new idea, as the teachers were accustomed to building curriculum around issues and topics that were meaningful to the students. In addition, the fact that the teachers assumed the responsibility for not having met the Bolivian children's needs no doubt positioned them to be comfortable in taking action to address the problem. In other words, a culture already in place at the school seems to have supported the success of the project.

What prospects for successfully addressing the educational needs of immigrant children are there if the culture of a school is unlike that of Escuela 10? It might be more difficult to mobilize an entire school that does not strongly value the need to build on the strengths and interests of children. A place to begin may be with individual teachers. Within schools of education in the United States, more attention is being paid to supporting preservice teachers in understanding the diversity they will meet in the classroom (Cochran-Smith, Davis, and Fries, 2004; Ladson-Billings, 1994, 1999). Developing this orientation may support them in changing the way they relate to children who are unlike them, by recognizing the value of being proactively inclusive of them within the classroom and seeing strengths in their families and communities that can be used in support of their academic achievement. Professional development is another way to tackle the issue of including all children and teaching them well. Yet, in our quest to teach "all" children, we must not forget that to do this well we must begin by teaching particular children. The story of Escuela 10 speaks to this point. It is a story that goes beyond learning about how to address the needs of Bolivian immigrant children living in Argentina. It speaks to much more global principles, such as school faculties working together to address the needs of particular children in their schools and the impact this can have on student participation and achievement.

Note

1. The school year in Argentina, as in other countries in the Southern Hemisphere begins in March and ends in December of a single calendar year.

References

Civil, M. (1998). "Parents as Resources for Mathematical Instruction." Paper presented at the ALM-5, Netherlands.

———. (2000). "Parents as Learners and Teachers of Mathematics: Toward a Two-Way Dialogue." Paper presented at the International Conference of Adults Learning Mathematics, Medford, MA.

Civil, M. and Bernier, E. (2004). "Parents as Intellectual Resources in Mathematics Education: Challenges and Possibilities." Paper presented at the Annual Meeting of the National Council of Teachers of Mathematics, Philadelphia.

Cochran-Smith, M., Davis, D., and Fries, K. (2004). "Multicultural Teacher Education: Research, Practice, and Policy." In J. Banks and Banks, C. A. M. (eds.), *Handbook of Research on Multicultural Education* (2nd ed.). San Francisco: Jossey-Bass. 931–975.

Foote, M. (2001). *Informal Talks with Argentine Elementary School Teachers.* Buenos Aires and Luján, Argentina.

———. (2006). "Supporting Teachers in Situating Children's Mathematical Thinking within Their Lived Experience." Unpublished manuscript, Madison, WI.

González, N., Andrade, R., Civil, M., and Moll, L. (2001). "Bridging Funds of Distributed Knowledge: Creating Zones of Practices in Mathematics." *Journal of Education for Students Placed at Risk*, 6(1,2), 115–132.

Gutiérrez, R. (2002). "Enabling the Practice of Mathematics Teachers in Context: Toward A New Equity Research Agenda." *Mathematical Thinking and Learning*, 4(2,3).

Ladson-Billings, G. (1994). "Who Will Teach Our Children? Preparing Teachers to Successfully Teach African American Students." In E. R. Hollins, King, Joyce E., Hayman, Warren C. (eds.), *Teaching Diverse Populations: Formulating A Knowledge Base.* Albany, NY: State University of New York Press. 129–142.

———. (1999). "Preparing Teachers for Diverse Student Populations: A Critical Race Theory Perspective." In A. Iran-Nejad and C. D. Pearson (eds.),

Review of Research in Education, 24, 211–247. Washington, D.C.: American Educational Research Association.

Moll, L., Amanti, C., Neff, D., and Gonzalez, N. (1992). "Funds of Knowledge for Teaching: A Qualitative Approach to Developing Strategic Connections between Homes and Classrooms." *Theory into Practice*, 31, 132–141.

Moll, L. and Gonzalez, N. (2004). "Engaging Life: A Funds-of-Knowledge Approach to Multicultural Education." In J. A. Banks and C. A. M. Banks (eds.), *Handbook of Research on Multicultural Education* (2nd ed.). San Francisco: Jossey-Bass. 699–715.

CHAPTER 8

Teaching against Threats to Democracy: Inclusive Pedagogies in Democratic Education

Diana E. Hess and Shannon Murto Wright

The idea that schools in democratic nations should teach toward the continuation of democracy, at a minimum, and ideally toward its positive transformation, is not new or novel. But while it receives a lot of rhetorical support, evidence suggests that many people do not really believe that this is a task for schools to take up with respect to their children, especially relative to other roles such as credentialing for further education and workplace preparation. Peter Levine, the executive director of the preeminent civic learning and engagement research center in the United States, argues that markets "pose special problems for *civic* education. The civic development of young people will be undervalued in any market system, unless we take deliberate and rather forceful efforts to change that pattern" (Levine, 2005). There is evidence to support his claim. When asked to rate whether preparing young people for democratic participation was a very important goal of schooling, just over 50 percent of adults in the United States agreed, while other goals, such as workplace preparation (which 64 percent rated as very important) and basic academic knowledge in reading, math, and science (80 percent agreed) received much more support (Campaign for the Civic Mission of Schools and Alliance for Representative Democracy, 2004).

In a time when the threats to democracy are numerous and powerful, the very possibility that schooling can play any kind of meaningful role in the creation, maintenance, or transformation of democracy can seem both idealistic and hopelessly naïve. Though there is reason to be suspicious of how broad and deep the support is for the role that schools can and should play toward the end of shaping and improving democracy among the general public, it is striking that in this time when democracy is increasingly challenged and challenging, it is still fairly common for teachers to talk with passion and fervor about themselves as democracy workers. In Hess's research in middle and high school social studies classes, she regularly encounters teachers who say they hope their practice not only shape how young people view democracy, but also help them develop the ability and desire to engage politically and civically. To be sure, these teachers see contemporary threats to democracy as barriers, but they also see them indicators of why it is so important for schools to take up this role in the first place and to act as a catalyst for action (Hess, 2002; Hess and Posselt, 2002; Hess and Ganzler, 2006).

Although it may be the case that the most significant challenge to democracy is that there are so many challenges, in this chapter the focus is on just two challenges and how inclusive pedagogies can help combat these challenges. The two challenges, while linked, have particularized dimensions with significant implications for schooling writ large and for democratic education in particular. The first challenge is the increasing disparity between the rich and the poor in the United States, which, as recent evidence illustrates, has extremely serious consequences for political and civic participation and social inclusion. The second challenge is the inability of many people to conceptualize political conflict as a social and political good and act in accordance. Related to this is the withdrawal from any meaningful notion of a "public" by creating private spaces and institutions marked by the ideological and economic homogeneity of those who inhabit them—both in the larger society and within schools.

Both of these challenges to democracy operate in society and also have particular manifestations inside numerous schools. There are ways that many schools are not interrupting these challenges to democracy—but instead reifying them and amplifying their power. However, that is not the case for all schools, or all teachers. This is where inclusive pedagogies can become powerful tools in the hands of skilled teachers who know how to implement them in the classroom. In fact, there are numerous examples of how schools and teachers are working to consciously and deliberately teach against the threats to democracy.

Our plan for the chapter is to address the two challenges by first explaining what they are outside of the schools, then explore how they operate inside schools, and finally describe how some schools and teachers are using inclusive pedagogies to teach against this challenge. While we focus here on U.S. democracy and schooling, recent work by Hess with teachers and professors in the Czech Republic, Northern Ireland, the Republic of Ireland, and England leads us to believe that some of issues we raise here are not unique to the United States alone, for this reason, our hope is that they have some relevance for readers in other nations too. We conclude with a caution about not overselling the impact of schools—not withstanding our belief that schooling really can influence the health of democracy. Nevertheless, either out of naivety, or sometimes purposeful distraction, there are people who place too much of an emphasis on the power that schooling has to address some of the ills that so desperately need attention.

Political Conflict as a Good

There is a long line of both theory and research to support the claim that democracy depends on and is strengthened by controversy (Dahl, 1998; Parker, 2003). This is so because in well-functioning democracies controversy *in practice* is a synonym for a fair hearing of multiple and competing views about important public questions that deserve attention. The only way to guard against the tyranny of the majority or the close-mindedness of the like-minded community is to work toward a political communication culture that is rich with difference through inclusion. One way to measure the health of a democracy, then, is to gauge whether people come to public life with an a priori appreciation for the foundational role that controversy must play in any reasonably healthy democracy and also a willingness to voice their own views and take seriously the views that differ from their own.

At the heart of the deliberative enterprise is the foundational belief that multiple perspectives are an asset—not a hindrance—to democratic thinking, participation, and governance. After all, if people agreed on what should be done about important problems, what would be the purpose of deliberation? In such a circumstance, it would be logical to move forward with putting into place the solution that enjoyed such widespread support. But we know that many of the public's problems do, in fact, involve controversy because there are fundamental disagreements about which problems deserve attention in the first place, what has caused the problem, how much of a problem it is *really*, for whom it is a problem, what are the strengths and weaknesses of proposed solutions, and the list goes on. In short, while there are some things that vast majorities of people agree on, it is also the case that there are many disagreements—as one would expect in democratic nations. In fact, the extent to which people disagree about important problems facing the community is not a flaw of democracy but rather a marker of how democratic a community is in practice.

Consequently, it is not surprising that many share a basic belief that when you bring together a group of young people in a classroom, disagreements will happen, and those disagreements are something to be cultivated, not suppressed. As a case in point, the United States Supreme Court 2003 ruling in *Grutter v. Bollinger* revolved around the question of whether the 14th Amendment's equal protection clause, which in almost all cases makes government decisions based on a person's race *verboten*, could be trumped by the *compelling interest* the state had in ensuring that students were exposed to differences as a part of their education (in this case, through affirmative action at the University of Michigan Law School). By a slim majority, the Court answered in the affirmative. Diversity, as the majority construed it, was not about redressing historic wrongs against people who had historically been discriminated against—the Court did not buy the claim that affirmative action in this case was a necessary form of redress—but instead, it was an educational asset because students learn more when they are in inclusive settings in which they are exposed to views that differ from their own. The Court's decision in the case is important because it represents—officially—the commonsense notion that if you are going to talk about hard problems, it is better to have more opinions represented at the table. This not only helps to ensure that the decision does not just represent narrow interests (a substantive good), but also that people do not feel that their views are kept from consideration as a matter of course (a procedural good).

The Democracy Divide as a Threat to Democracy

One major threat to a healthy, deliberative democracy is the growing "democracy divide." New evidence suggests that democratic education opportunities are distributed in ways that privilege wealthier students. A key series of studies by Joseph Kahne and Ellen Middaugh (2008) have demonstrated that elements of high-quality democratic education that help develop students' ability and desire to engage politically and civically are much less likely to be experienced by students who are poor, African American, or Latino. To draw one illustrative example from their studies, students in classes with higher average socioeconomic status (SES) levels are 1.42 times more likely to report participating in debates or panel discussions in their social studies courses than students in classes with lower average SES levels.

The issue here is serious, indeed, as there is an increasing disparity between the rich and the poor in the United States. This gap recently was highlighted by data released by the Congressional Budget Office (2007), which revealed that between 1979 and 2005 the average household income for the lowest fifth of U.S. households increased by only 6 percent. For the highest fifth of U.S. households, the average income increased by 80 percent; for the top 1 percent of U.S. households, the increase was an astonishing 224 percent. This data clearly shows that the gap between high SES households and low SES households in the United States has grown at an astonishing rate in recent decades.

As a result, not only is there a growing income gap between the rich and the poor in the United States, but as Kahne and Middaugh (2008) so clearly show, students in classes with a lower average SES levels receive fewer opportunities to engage in the kind of high quality democratic education opportunities that might lead to greater political engagement and involvement later in life. Just as the growing income gap diminishes the economic influence of the nation's lowest SES citizens, the democracy divide threatens to diminish the political influence of the same group through a reduced number of opportunities (as compared to higher SES groups) to participate in activities that might increase their political efficacy and engagement. The overall result of receiving fewer opportunities to engage in high quality democratic education is that citizens with lower SES are more likely to be disconnected from the democratic process than their higher SES peers.

The growing democracy divide in schools also may be influencing a growing democracy divide among young voters in the United States. Recent evidence indicates that young people with lower educational attainment are not "aging into" political participation in the way many of their predecessors did. In fact, on practically every measure of political and civic activity (e.g., voting, attending political meetings, buying/boycotting?) paying attention to the news, deliberating with others about public issues), young people with a high school education participate at a much higher level than their peers who did not graduate from high school. The rate for college attendees is higher still. For example, during the early 2008 presidential primary season, one in four eligible young people with college experience voted on Super Tuesday, while only one in fourteen among the noncollege youth voted in the U.S. primaries (Marcelo and Kirby, 2008).

This relationship between education and political participation is not a new trend, but perhaps never before has educational attainment been so predictive of whether young people will participate in politics. In a nation where unemployment is almost three times higher than the voting rate of noncollege attending young people in the most recent electoral primaries, it should come as no surprise that trends in the economy generally, the quality of educational opportunities afforded to young people, and the participation of young people in democracy are all intrinsically connected. These examples demonstrate that many citizens (particularly those with low SES or low educational attainment) receive fewer opportunities to engage in the type of high quality democratic education that might lead them to become more active in the public sphere.

Pedagogies against the Democracy Divide

The threat of the democracy divide, marked by relatively few opportunities to engage in high quality democratic education for students of low SES or low educational attainment, has not gone unnoticed among teachers. These teachers realize that high quality democratic education needs to be distributed evenly to students of all backgrounds, and they realize that students of all backgrounds can benefit from exposure to political interaction with students whose backgrounds differ from theirs.

An illustrative example of teachers acting with these goals in mind comes from an ongoing study conducted by Hess that examines democratic education in high school courses (Hess and Ganzler, 2006). One school's required senior Government class stands out as an instructive example, because the manner in which the course is conducted reveals some of the ways in which inclusive pedagogies potentially can minimize the deleterious effects of this threat to democracy.

The first way in which this school works to ensure that high quality democratic education is equally distributed and that students are exposed to a diversity of political viewpoints is to require a senior-level, untracked Government course. As one teacher noted of the Social Studies department's decision to refrain from tracking Social Studies courses, "one of the big beliefs of our former department chairman and the department chairman before him was that...we are all in the country together and we are all in this government together and so we probably should learn about all of it together." As a result of this strongly held belief within the department, every senior student in the high school takes a similar Government class, although "Sheltered" classes exist for English Language Learners.

However, the school maintains its commitment to the equal distribution of high quality democratic education and to exposing students to a wide range of political viewpoints through school-wide simulations. Students in every senior Government class, including in the Sheltered classes, participate equally in senior class-wide simulations with their peers. The teacher of the Sheltered Government class notes a personal belief that English Language Learners should "experience the things the mainstream kids do...for these kids to be part of the simulation, take on leadership roles, have their own legislation, get up [to] speed, is really powerful." This teacher highlights one of the important reasons to involve all students in high quality democratic education, which is the impact it has on students who have had less exposure in the past to this type of education.

Another important reason to involve all students in high quality democratic education is the impact that such involvement can have on students' views of differing political ideas. Of course, the goal is not to *change* students' viewpoints, but rather to encourage them to become more tolerant and more understanding of the views of others. As one student in the Government class remarked, hearing the opinions of others "opens my mind, too, to kind of think, well, if I thought that way...it gives you both aspects." This exposure to a diversity of viewpoints is one way to counteract the tendency toward extremism that can occur when citizens huddle in communities of sameness.

Another approach is to make sure students understand the threats to democracy, that is, to utilize a curriculum that focuses explicitly on exposing and discussing the deep fissures in U.S. democracy, as well as publicly discussing different viewpoints on how to address these fissures. One of the teachers in the study commented on how the Social Studies department addressed some of these fissures: "one of our big pushes in the Social Studies department has been to use particularly American History and American Government as a way to bring economic and racial issues to the forefront as opposed to ignoring them. And our curriculum has been designed around that." By bringing these issues to the forefront instead of allowing them to linger in the background, students and teachers can begin to confront the issues that threaten to divide democracies within the school environment, a process that hopefully will begin to develop a comfort level for students to confront these issues in society outside of schools.

Conflict Avoidance as a
Threat to Democracy

The second threat to democracy is the growing trend toward conflict avoidance in society and in schools. Conflict aversion threatens democracy both by increasing the likelihood that fewer voices will be heard in the discussion of issues and by narrowing the range of options that are brought forward as possible solutions to pressing public problems.

In fact, in the course of real democratic practice, it is not clear whether the claims of the substantive and procedural goods that are embodied in a diverse and inclusive political community really pay off with respect to actual participation. Of late, scholars have shown empirically that there is an inverse relationship between racial (Putnam, 2007) and ideological diversity (Mutz, 2006) in a political community and in political and civic participation. What is interesting to note is that both of these highly regarded scholars report their findings with something akin to dread; it is clear that their belief in the deliberative ideal—that a diverse community is good and will cause political participation—will not go down without a fight. Diana Mutz (2006) argues that her findings that show the extent to which engagement in "cross-cutting political talk" seems to usher in tolerance and decrease political engagement is not a reality that we should get used to—but a problem to be remedied.

Similarly, Hibbing and Theiss-Morse (2002) have identified the aversion that many adults in the United States have toward political conflict and the relationship that exists between that aversion and the unwillingness to participate in the political realm. Like Mutz (2006), they are clearly distressed by the problem they have identified and urge the schools to act as an intervention by teaching young people that conflict is natural, not something to be shunned. David Campbell (2005), in his analysis of the IEA Civic Education Study data, makes the leap from racial diversity to ideological diversity and shows an inverse relationship in schools between the diversity in a classroom and the amount of discussion students report about issues.[1] To conclude, it is hard to read the literature about the relationship between ideological diversity and democracy without coming to the conclusion that there are many more advocates for its intrinsic and theoretical connections than there is actual evidence for the relationship working that way in practice.

Certainly, Hibbing and Theiss-Morse are not the only scholars documenting the low level of political engagement among citizens of the United States. Their study (Hibbing and Theiss-Morse, 2002) is particularly compelling, however, because it pinpoints the fact that people in the United States generally *like* conflict and controversy (witness the addiction to viewing competitive sports), but they dislike conflict and controversy when it is related to politics, policy issues, and governance. Consequently, the aversion to conflict and controversy causes low levels of political engagement. Moreover, it dampens the appetite for a wide range of political views, which may account for why visitors to the United States from other democracies so frequently comment on the relatively narrow range of political views available in newspapers and television news compared to what they are used to. People in the United States do not demand wide diversity in their political news climate—and, not surprisingly, they do not receive it either.

Another symptom of the aversion of many to political conflict is how few people actually engage in political discussion. In a study of adults in six communities in Britain and the United States, Conover, Searing, and Crewe (2002) found that 30 percent of the sample in the United States and 50 percent in Britain are "silent citizens." That is, a large percentage of each nation's respondents discuss issues in private only. Virtually no one discussed issues in the public sphere only, and a mere 18 percent of U.S. citizens and 9 percent of British citizens reported speaking in both contexts. Further, Conover, Searing, and Crewe (2002) suggested that these discussions are often marred by inequality and a lack of analysis and critique. The aversion to political conflict and controversy is a threat to democracy, because it then becomes likely that only fewer voices will be heard and also narrows the range of options that will be considered as possible policy solutions to pressing problems.

A "Private" Democracy

Another aspect of conflict avoidance is the increasing withdrawal from any meaningful notion of a "public" by creating private spaces and institutions marked by the ideological and economic homogeneity of those who inhabit them. In *The Big Sort*, Bill Bishop demonstrates that "as Americans have moved over the past three decades, they have clustered in communities of sameness, among people with similar ways of life, beliefs, and in the end, politics" (2008, 5). Consequently, legislative districts are being ideologically gerrymandered to an extent that was unheard of in the past. For example, in 2004 more than 50 percent of people in the United States lived in a county where one of the presidential candidates won by a landslide—compared to only 26 percent in 1976 (Bishop, 2008).

George Stephanopoulos and ABC News (2006, June 30) found a similar trend in two experiments examining how living in communities of sameness can shape public life. In the first study, conducted by Cass Sunstein, the results demonstrated that like-minded people adopted more extreme positions when grouped together. In the second study, Diana Mutz found that television viewers were very likely to misunderstand people who held viewpoints that opposed their own when they watched people shouting at each other (an increasingly common aspect of national news commentary shows).

The implication of both studies taken together is that the trend toward communities of sameness—a trend that is only accelerating through the Internet, where self-selection defines web "communities" such as message boards and blogs—can exacerbate the extremism of people's political views. Although this extremism might be combated by the exposure to differing views through national media like television and the Internet, the tendency toward "shouting heads" on television and the self-selection aspect of the Internet prevent these media from serving as a balancing force to the increasing extremism that can result from communities of sameness.

CONFLICT AVOIDANCE IN SCHOOLS

The conflict aversion that occurs in society as a whole also finds its way into society's schools. While there are a numbers of ways in which conflict aversion manifests itself in schools, three of them seem to be particularly problematic: the narrowing of ideas in the curriculum, the lack of aptitude that students have in dealing with ideas that challenge their own, and the infrequency in which many students experience and are explicitly taught how to engage in productive discussions that focus on conflictual content.[2] These problems are not new, nor are they unique to the United States.

Numerous researchers have documented how textbooks marketed in the United States tell one particular story—and rarely invite students to consider the reality that there are typically multiple and competing stories that are worth consideration (e.g., Zimmerman, 2002). In a study directed by Hess that analyzed the portrayal of 9/11 and terrorism in nine best-selling social studies textbooks, we found almost a mind-numbing sameness within and across the books. Few of them gave even a nod to the reality of the controversies that exist in the political world outside of school, such as what terrorism means and what are the best responses to terrorism (Hess, Stoddard, and Murto, 2008). Moreover, we found that most textbooks fail to ask students to think very deeply about the content. Of the ninety-five questions related to 9/11 and terrorism in these nine textbooks, only forty asked higher-order thinking questions with "open" answers—that is, answers that were not already determined in the text; further, only fourteen of these forty questions provided enough additional information (a scaffold, so to speak) for students to construct an answer.

A related problem is created by the fact that students represent the larger political culture that shuns political conflict and controversy; as a result, few come into classes with the skills or aptitude to engage thoughtfully and critically in discussions and activities that depend on multiple positions, perspectives, and views. In fact, research shows that many students believe that as long as something is their opinion, it should not be challenged (Hess and Posselt, 2002).

Finally, although discussion is often advocated as a key vehicle for the study of controversial issues, fairly large-scale observational studies involving middle and high school social studies classes report virtually no classroom discussion (Nystrand, Gamoran, and Carbonara, 1998). Moreover, the discussions that do occur do not focus on controversial issues (Kahne, Rodriguez, Smith, and Thiede, 2000).

Martin Nystrand and his colleagues (1998) analyzed discourse in 106 middle and high school social studies classes, each of which was observed four times throughout the school year. Nystrand, Gamoran, and Carbonara report that "despite considerable lip service among teachers to 'discussion,' we found little discussion in any classes" (1998, 36). To account for this, Nystrand explains that teachers typically conflate some form of recitation (such as the familiar IRE pattern of teacher-initiated question, student response, and teacher evaluation) with discussion. Defining discussion as the free exchange of information among three or more participants (which may include the teacher), the researchers note that they did observe some discussions, but that 90.33 percent of what they viewed involved no discussion, and the remaining discussion time was brief: on average, 42 seconds per class in eighth-grade classes and 31.2 seconds per class in ninth-grade classes.

Kahne and his colleagues (2000) share a similar report in their observations of 135 middle and high school social studies classes in the Chicago Public Schools, finding that controversial issues receive scant attention. In over 80 percent of the classes observed, there was no mention of a social problem, and even when problems were mentioned, there was rarely any discussion of possible solutions, connections to modern times, or action.

These factors—the narrowing of ideas in the curriculum, the lack of aptitude that students have in dealing with ideas that challenge their own, and the infrequency in which many students experience and are explicitly taught how to engage in productive discussions that focus on conflictual content—are key attributes of the larger pattern of conflict aversion that currently exists both in society as a whole and in schools. This conflict aversion threatens democracy both by increasing the likelihood that fewer voices will be heard in the discussion of issues and by narrowing the range of options that are brought forward as possible solutions to pressing public problems.

PEDAGOGIES AGAINST CONFLICT AVOIDANCE

There are teachers who are quite aware of how conflict aversion—both outside and inside of school—is a reality that needs to be addressed within schools. These teachers characterize students' aversion to conflict as a misconception that must be purposely and carefully corrected. They are not interested in shaping students' political views per se, but are interested in ensuring that students embrace political controversy as a necessary and helpful reality in a functioning democracy (Hess, 2002; Hess and Posselt, 2002; Hess and Ganzler, 2007).

Many teachers work toward putting these views into practice by adopting and enacting a particular type of democratic education that centers on creating a more inclusive classroom by enabling their students to tackle the extremely controversial issues that animate contemporary political debates in the United States. Their method is to provide a forum for students to develop an understanding of issues, including weighing various perspectives on how they should be resolved, forming and expressing their views on the issues, and having those same views challenged by their classmates. The goal is to kindle student interest in the controversies, with the hope that this engagement may spark other forms of political engagement. Toward this end, the teachers want their students to develop an appreciation for the necessity of political conflict in a democracy and a willingness to engage with others around authentic issues, especially with people whose views differ from their own.

The barriers to this kind of pedagogical practice are numerous and high; most notably, many students are not used to hearing multiple and competing perspectives and sometimes mistake them for views they label as suspiciously out of the mainstream. To combat this, the teachers work to normalize political conflict by activating the ideological diversity that exists in their midst. This is done intentionally because the teachers see the ways in which political controversies are talked about in the world outside of school as a threat to democracy. They do not want to reinscribe these practices in their classrooms; on the contrary, they want to teach against the threat—to transform the landscape of the discussion of political controversies—by creating an inclusive environment in which students feel both welcome and free to express their thoughts on important issues. This is an extremely challenging goal that is much easier to conceptualize and yearn for than to put into practice.

However, there are key elements of the pedagogies used by teachers participating in an ongoing study of democratic education in high school courses that are worth explicating, because they make clear the contours of this threat to democracy and the ways in which inclusive pedagogies have the potential to minimize their negative effects.[3] While it is clear that teaching students to discuss controversial issues is an extremely complex enterprise that can be implemented successfully in a variety of different ways, some similarities stand out among teachers who achieve this goal.

First, skillful-discussion teachers recognize that many of their students do not come into their classes already possessing the background knowledge, communicative skills, and deliberative dispositions necessary to participate effectively in thoughtful civic discussions of highly controversial issues. Consequently, simply providing students with the opportunity to engage in such discussions is not sufficient, especially if the goal is to reach *all* students, as opposed to simply providing a forum for students who are already proficient discussants. Instead, teachers who are unusually good at this form of teaching carefully structure instruction to explicitly teach students the skills they need, such as how to ask clarifying questions, how to use different kinds of evidence, and how to use reasoning to back up or probe a claim.

Second, skillful-discussion teachers are extremely attentive to creating a classroom climate that encourages the airing of multiple and competing views and the participation of all students. Such teachers recognize that peer relations within a class are undoubtedly going to influence students' willingness to participate, as well as the nature and tone of the interaction. While there are many aspects of peer interactions teachers cannot control, they can ensure that within-class grouping does not reify existing power hierarchies among students, that students learn and use one another's names, and that students are not allowed to engage in the taunting and/or personal putdowns that unfortunately characterize some of the political talk we see in the world outside of school.

Third, skillful-discussion teachers do not aim for spontaneous discussions because they recognize that these rarely occur, and when they do, participation is typically minimal. Instead, discussions are carefully planned, students learn background information so they feel like they know enough to participate, and a particular model or form of discussion is used that aligns with the teachers' goals. These discussions are not "free for alls" that generate more heat than light. While the best discussions of issues are still lively and highly engaging, they are also organized and civil.

Although these teachers may vary greatly in the nuances of their approaches, the basic elements of explicitly teaching discussion skills, creating a classroom climate supportive of multiple and conflicting viewpoints, and planning and preparing for discussions cut across their varied techniques, offering a good starting point for teachers to examine their own practice of implementing inclusive pedagogies through the discussion of controversial issues in the classroom.

CONCLUSION

Notwithstanding the efforts that many schools and teachers are making to address the democracy divide, it is also clear that there are real limits to what schools can do—which is why it is dangerous for educators to believe that education *alone* is the key to improving democracy—schools can reduce the threat, but they are not going to narrow the widening gaps between the rich and poor in the United States. We shouldn't be shy about admitting how deep and difficult these problems are and how much of a threat they are to democracy. At the same time, schools can make a difference, although we should not kid ourselves and expect schools to transform society. So, what can and what should educators do?

This chapter has highlighted several approaches that teachers can take, and have taken, to work against some of the threats to democracy, that is, conflict avoidance and the growing democracy divide. It is our hope that these approaches can provide guidance for teachers seeking to implement strategies to combat the threats to democracy, or to evaluate and refine their own practices for teachers who already utilize such strategies.

The challenge is great, and the stakes are high. Although high quality democratic education is by no means the only answer to these difficult challenges to democracy, it is a place to start.

NOTES

1. The IEA Civic Education Study is a school-based survey on civic education administered to a representative sample of students in

twenty-eight countries, including the United States. For more information, see http://www.wam.umd.edu/~jtpurta/.

2. As a point of clarification, through "narrowing the curriculum," I refer to both the *material* curriculum (the range of courses offered; the content of textbooks, tests, and other sources that communicate the "knowledge of most worth") and the curriculum enacted through the nature of the intellectual work that students are asked to do with the materials.

3. The study referred to here is a four year longitudinal project that includes 21 schools, 1200 students, and 35 teachers in three states in the United States. See Hess and Ganzler (2006) for more details about its design.

References

Bishop, B. (2008). *The Big Sort: Why the Clustering of Like-Minded America Is Tearing Us Apart*. New York: Houghton Mifflin.

Campaign for the Civic Mission of Schools and Alliance for Representative Democracy. (2004). *From Classroom to Citizen: American Attitudes on Civic Education* [Electronic version]. Calabasas: Center for Civic Education. Retrieved from http://www.civicmissionofschools.org/site/resources/keyreports.html.

Campbell, D. E. (2005). "Voice in the Classroom: How An Open Classroom Environment Facilitates Adolescents' Civic Development." *CIRCLE Working Paper* 28. Retrieved from http://www.nd.edu/~dcampbe4/voice.pdf.

Congressional Budget Office. (2007). "Historical Effective Federal Tax Rates: 1979 to 2005." Washington, DC: Congressional Budget Office. Retrieved May 26, 2008, from http://www.cbo.gov/doc.cfm?index=8885.

Conover, P. J., Searing, D. D., and Crewe, I. M. (2002). "The Deliberation Potential of Political Discussion." *British Journal of Political Science*, 32, 21–62.

Dahl, R. A. (1998). *On Democracy*. New Haven: Yale University Press.

Grutter *v.* Bollinger, 539 U.S. 306 (2003).

Hess, D. (2002). "Discussing Controversial Public Issues in Secondary Social Studies Classrooms: Learning from Skilled Teachers." *Theory and Research in Social Education*, 30(1), 10–41.

Hess, D. and Ganzler, L. (2006). "How the Deliberation of Controversial Issues in High School Courses Influences Students' Views on Political Engagement." Presentation at the Annual Meeting of the American Educational Research Association, San Francisco, CA.

Hess, D. and Posselt, J. (2002). "How High School Students Experience and Learn from the Discussion of Controversial Public Issues." *Journal of Curriculum & Supervision*, 17(4), 83–314.

Hess, D., Stoddard, J., and Murto, S. (2008). "Examining the Treatment of 9/11 and Terrorism in High School Textbooks." In J. S. Bixby and J. L.

Pace (eds.), *Educating Democratic Citizens in Troubled Times: Qualitative Studies of Current Efforts*. Albany, NY: SUNY Press.

Hibbing, J. R. and Theiss-Morse, E. (2002). *Stealth Democracy: Americans' Beliefs about How Goverment Should Work*. New York: Cambridge University Press.

Kahne, J. and Middaugh, E. (2008). Democracy for Some: The Civic Opportunity Gap in High School. *CIRCLE Working Paper* 59. Retrieved May 26, 2008, from http://www.civicyouth.org/?p=278.

Kahne, J., Rodriguez, M., Smith, B., and Thiede, K. (2000). "Developing Citizens for Democracy? Assessing Opportunities to Learn in Chicago's Social Studies Classrooms." *Theory and Research in Social Education*, 28(3), 311–338.

Levine, P. (2005, December 15). *Why Schools and Colleges Often Overlook Civic Development*. Retrieved May 26, 2008, from Peter Levine Homepage: http://www.peterlevine.ws/mt/archives/2005/12/why-schools-and.html.

———. (2007). *The Future of Democracy: Developing the Next Generation of American Citizens*. Medford, MA: Tufts University Press.

Marcelo, K. B. and Kirby, E. H. (2008). "The Youth Vote in the 2008 Super Tuesday States: Alabama, Arizona, Arkansas, California, Connecticut, Georgia, Illinois, Massachusetts, Missouri, New Jersey, New York, Oklahoma, Tennessee, and Utah." *Circle Fact Sheet*. Retrieved May 26, 2008, from http://www.civicyouth.org/PopUps/FactSheets/FS08_supertuesday_exitpolls.pdf.

Mutz, D. C. (2006). *Hearing the Other Side: Deliberative versus Participatory Democracy*. New York: Cambridge University Press.

Nystrand, M., Gamoran, A., and Carbonara, W. (1998). *Towards An Ecology of Learning: The Case of Classroom Discourse and Its Effects on Writing in High School English and Social Studies*. Albany: National Research Center on English Learning and Achievement.

Parker, W. C. (2003). *Teaching Democracy: Unity and Diversity in Public Life*. New York: Teacher's College Press.

———. (2006, March 9). "Public Schools are Hotbeds of Democracy." *Seattle Post Intelligencer*, B7.

Putnam, R. D. (2007). "E pluribus unum: Diversity and Community in the Twenty-First Century: The 2006 Johan Skytte Prize Lecture." *Scandinavian Political Studies*, 30(2), 137–174.

Stephanopoulos, G. and ABC News. (2006, June 30). "A Country Divided: Examining the State of Our Union." *20/20*. Retrieved May 26, 2008, from http://abcnews.go.com/2020/Story?id=2140483&page=1.

Zimmerman, J. (2002). *Whose America: Culture Wars in the Public Schools*. Cambridge, MA: Harvard University Press.

PART III

Toward Inclusionary Teacher Education and Professional Development

CHAPTER 9

Pedagogies of Inclusion in Teacher Education: Global Perspectives

Christine E. Sleeter

Educators around the world increasingly see teacher education as crucial to developing pedagogies of inclusion, particularly as student populations diversify. While neoliberalism greatly contributes to the growing diversity of students for whom teachers need preparation, however, it is also shrinking public resources for serving those same populations while simultaneously constraining the work of teachers and teacher educators.

Expansion of global capitalism has prompted large-scale migrations of peoples, repopulating schools and communities on a scale not seen before (Suárez-Orozco, 2001). Around the world, in countries as different from each other as Hong Kong (Yuen, 2002), Greece (Vidali and Adams, 2006), the United States (Suárez-Orozco, 2001), and Spain (Soriano, 2008), schools are struggling, often for the first time, with the question of how to respond to newly arrived students. For example, Korea has recently attracted many foreign workers, and a significant proportion of its marriages are now international; bicultural children entering elementary schools confound their unprepared teachers (Uhn, 2007). At the same time, it is also imperative that schools improve education for historically marginalized communities in their own countries so that young people are not locked out of economic and political participation. Examples of such communities range from the Mapuche in Chile (Quilaqueo, 2006) and the Maori in New Zealand (Bishop and Berryman, 2006) to African Americans in the United States (Ladson-Billings, 2006), Roma in Europe (Katz, 2005), and Dalits in India (Thorat, 1999). To complicate matters further, nation-building in postcolonial contexts has fostered debates and struggles over equity, justice, and national identity, often in the midst of tremendous ethnic, religious, and linguistic diversity, historical complexity, and wealth coexisting with deep poverty (Meuleman, 2006).

While contributing to teacher education's challenges, however, neoliberalism is also constricting teacher education. Cuts in public expenditures imply that in many nations, teacher education has become shorter. For example, preservice programs in the United States had gradually lengthened between the 1970s and the early 1990s when general studies and clinical experiences expanded and programs developed field experiences and coursework to address changes in schools. In the early 1990s, however, due to cuts in public expenditures on higher education and competition from private vendors offering very short certification programs, average program length began to shrink, not only in the United States but in other countries as well (Feistritzer, 1999; Lyall and Sell, 2006; Openshaw, 1999). In addition, teaching in many areas of the world is becoming more technocratic as education is being defined as preparation for work; curricula are being oriented toward corporate needs and the work of teachers are being defined accordingly (Compton and Weiner, 2008; Puiggrós, 1997).

Considering this broad context, this chapter will examine what teacher education can accomplish. Drawing on examples of programs in various countries, I will argue that teacher education stands to benefit by engaging with its local communities, both as a way of preparing teachers for diversity and also as a way or pushing back against neoliberalism.

Constructing Pedagogies of Inclusion in Teacher Education

Pedagogies of inclusion in teacher education rest on three pillars. One pillar—*the university*—encompasses professional knowledge and theoretical grounding for inclusive curriculum and practice. For example, Darling-Hammond and Bransford (2005) conceptualized professional knowledge in three overlapping domains: knowledge of learners, which includes the learning process and how learning is prompted, guided, and transferred; the child developmental process; and the language development process, including development of linguistic skills in more than one language. Knowledge of curriculum includes designing and planning curriculum, as well as envisioning it in relationship to broad societal goals for the school. Knowledge of teaching encompasses a range of knowledge and skills for organizing learning, teaching subject matter, building teaching processes on cultural repertoires, linguistic skills, and other varying abilities related to students, assessing learning to guide instruction, managing the classroom, and collaborating with other professionals and parents. Zeichner (1996) has synthesized generally agreed-upon dimensions of knowledge for multicultural teacher preparation: clarification of teacher candidates' ethnic and cultural self-identities; self-examination of ethnocentrism; dynamics of prejudice and racism, and how teachers can address these; dynamics of privilege and economic oppression, and how schools contribute to these; multicultural curriculum development; the promise and potential dangers of learning styles; relationships between language, culture, and learning; culturally appropriate teaching and assessments; exposure to examples of successful teaching; and experiences in communities and schools.

A second pillar underlying pedagogies of inclusion—*the classroom*—includes guided practice working with everyday realities and complexities of diversity and inequity in the context of teaching. Model teacher preparation programs feature extended fieldwork in classrooms that serve diverse students, are built around close partnerships with universities that often locate coursework in the schools, and include active mentoring of teacher candidates by exemplary teachers (Rubenstein, 2007). However, one of the challenges is preparing teachers to transform and not simply replicate prevailing practices. According to Feiman-Nemser and Buchmann (1985), classroom experience has three limitations: familiarity that reinforces taking prevailing patterns for granted, divergent demands of universities and classrooms that prompt teachers to bifurcate rather than synthesize what they learn in each context, and incorporation of novices into already running systems rather than into classrooms that serve as labs for experimentation.

While the two pillars discussed above can be designed to promote progressive and inclusive pedagogies, they very often reinforce a standardized view of children, curriculum, and pedagogy that grows out of professional conceptions of "best practices," cultural homogeneity among classroom teachers, and the press of school bureaucracies. A third pillar of teacher education—*communities in which schools are situated*—offers potential to transcend universalized and standardized concepts of students, teaching, and learning. Community contexts tend to be absent from most discussions on teacher education, but I believe they are fundamental to pedagogies of inclusion. This argument was developed very thoughtfully four decades ago by the Study Commission on Undergraduate Education and the Education of Teachers (1976).

The Study Commission envisioned community as central to the preparation of teachers for at least two reasons. First, teachers will "have to know the language and culture of the children and youth they teach" (24). This is not simply a pedagogical necessity, as court cases and treaties in the United States guarantee the rights of historically marginalized communities to maintain their culture and language. Teachers, therefore, must be able to respond to and support the local community culture and language. Second, communities are a fundamental unit of social organization that schools can help to revitalize. Human welfare depends on close connections between individuals and local communities. As mass societies that are exacerbated by neoliberalism exert pressures that weaken communities, psychological welfare of citizens is diminished. By engaging schools in the life of local communities, not only can teaching become more meaningful to students, but the community itself can be strengthened. For these reasons, the Study Commission envisioned communities not only as a context for teacher education but also as active partners in deciding the nature of education for their children and the preparation of teachers to work with them.

In this chapter, I highlight several programs in different national contexts that address all three pillars: university, classroom, and communities. I focus particularly on ways in which teacher education programs (at both preservice and professional development levels) work with and collaborate with historically marginalized communities. I have organized this discussion around somewhat different ways of building inclusion: through school-community dialog, through student voice, through community-based service learning, and through political consciousness-raising. To be sure, there is overlap among these four sections, but this organizational structure provides a way of highlighting different emphases of the community pillar of teacher education.

Inclusion through School-Community Dialog

Teacher education can be located at the nexus of school-community dialog. In the two examples below—one from Chile and one from Spain—faculty members in teacher education facilitates bridge-building between schools and communities. In this dialogical context, teachers or teacher candidates learn to work with communities and translate community knowledge into the classroom.

A teacher preparation project in Chile—Pedagogía Básica Intercultural en un Contexto Mapuche (Elementary Intercultural Education in a Mapuche Context)—illustrates bicultural dialog as a basis for rethinking classroom practice. According to Quilaqueo (2007), a central problem in the preparation of teachers for indigenous Mapuche communities is that most teachers are steeped in Western knowledge and worldviews rather than those of the indigenous peoples of Chile. Too often, if teachers learn about indigenous knowledge at all, it takes the form of teaching techniques or social activities to include in the classroom. A much deeper concern is that Mapuche and non-Mapuche people approach the world, and each other, through deep cultural frames of reference that non-Mapuche teachers typically do not recognize, but that Mapuche adults have learned to navigate. The relationship between Mapuche and non-Mapuche knowledge is also hierarchical, with Western knowledge positioned as more scientific and modern. Typically, formal teacher education embodies a scientific and theoretical Western perspective, even when presenting information about "Others," implicitly giving secondary status to Mapuche knowledge.

To address this problem, Quilaqueo and his colleagues have been figuring out ways of engaging teachers in dialog with Mapuche communities in order to reconstruct classroom practice. They are presently researching the epistemology of Mapuche knowledge and creating a third space in which Mapuche and non-Mapuche can meet. The teacher education program helps both Mapuche and non-Mapuche teacher candidates who are interested in such preparation to learn to develop dialogical relationships with each other, as well as with members of the Mapuche community. Quilaqueo points out that everyone comes to teacher education from a cultural frame of reference; the challenge is making that frame of reference explicit and learning to engage with people whose cultural frame is different from one's own. In the case of this program, some of the faculty members and teacher candidates are Mapuche; the matter is not one of non-Mapuche people training other non-Mapuche people to work in Mapuche settings, but rather one of learning to establish intercultural collaboration throughout the entire program. The program is fairly new. At this stage of its research, Quilaqueo believes that structuring teacher education around ongoing dialog and collaboration in which the community, the teacher preparation program, and teacher candidates become interdependent has great promise.

A project in the south of Spain is developing teachers' ability to work with immigrant students by connecting schools with immigrant community associations (Soriano, 2008). In the late 1990s, Soriano and her colleagues investigated teachers' concerns about teaching newly arrived immigrant students. They found that teachers were unable to communicate with immigrant students and their families, were unfamiliar with the cultures of immigrants, lacked relationships with immigrant families, and lacked time as well as training to directly address these problems. Spaniards (including teachers) also assumed that immigrants brought values that conflict with those of the native Spaniards.

In a study funded by the government of Andalucía, Soriano then investigated the extent to which immigrant values do actually differ from Spanish values, identifying areas of overlap (such as shared value for the family). She reasoned that communication and collaboration between schools and immigrant communities could begin with recognition of shared values. Interviews were conducted with representatives from sixteen immigrant associations (e.g., one association representing immigrants from Mali and three associations representing immigrants from Morocco) and with teachers in primary and secondary schools in regions that serve students represented by the immigrant associations. Soriano (2008) found that both teachers and immigrant associations expressed similar reasons to explain why it would be beneficial to work together in the school.

For the most part, Soriano's research found teachers open to the idea of schools mediating between cultural groups, but they were unprepared to do so by themselves (Ejbari and Soriano, 2006). Soriano and her colleagues have been working with immigrant associations to build dialog and collaboration between schools, teachers, and immigrant communities, and, in the process, provide a form of teacher professional development resulting from dialog and collaboration. Immigrant community associations have been particularly beneficial partners with schools because they are able to interface between immigrant families and Spanish institutions, and they also have a commitment to working out solutions to problems that immigrants face. Teachers are learning to recognize that immigrant associations bring a wealth of knowledge that can help teachers learn how to teach diverse populations in the classroom and how to mediate conflicts between native Spanish students and immigrant students (Soriano and Ejbari, 2006). With university facilitation, teachers have begun to work with immigrant associations in making curricular and pedagogical changes in the classroom. For teachers, collaborating with immigrant associations is not only helping them to learn to communicate with and teach their immigrant students but also learn to improve the attitudes of native Spanish students toward the immigrants (Soriano, 2008).

Inclusion through Students' Voice

The two programs above work toward inclusion by engaging teachers in dialog with adults from communities that are culturally different from their own. A different approach is to ask students from a historically marginalized community to identify what works best for them in school, and then this data can be used as the basis for teacher education. A project in New Zealand, Te Kotahitanga, has been designed in this way (Bishop, Berryman, Tiakiwai, and Richardson, 2006). This project currently involves professional development of practicing teachers in thirty-three secondary schools; it is projected to expand to encompass elementary education and also preservice teacher education.

The project began with the Maori community's concern about poor academic achievement of Maori students. As one avenue to address this problem, in 2001 a Maori research team gathered narratives from Maori high school students, their family members, their teachers, and school administrators regarding the main influences that limit as well as support Maori student achievement. A striking pattern that emerged in the narratives was that while teachers mainly described the students and their families as contributing to Maori students' achievement problems (using a cultural deficit framework), the students talked mainly about relationships with teachers as central to helping them learn, and also as the factor that was lacking in most classrooms (Bishop and Berryman, 2006).

Based on an analysis of the student narratives, the research team constructed an effective teaching profile that represented the kind of pedagogy that would work for the Maori students. The team posited that effective teachers of Maori students "positively and vehemently reject deficit theorizing as a means of explaining Maori students' educational achievement levels" and know how to "bring about change in Maori students' educational achievement and are professionally committed to doing so." The effective teaching profile includes (1) caring about students as culturally located beings, (2) caring for the academic performance of students, (3) creating a secure, well-managed learning environment, (4) engaging in effective teaching interactions with Maori students *as* Maori, (5) promoting effective teaching interactions and relationships, and (6) monitoring and reflecting on outcomes that lead to achievement by Maori students (Bishop and Berryman, 2006, 273).

A professional development program was then constructed on the basis of this profile. Teachers read the narratives to find out how Maori students and their families view their schooling, and how teachers tend to see it. For many teachers, the very unsettling experience of seeing themselves reflected in the narratives was enough to prompt them to want to make changes. Reflecting on the narratives is followed by ongoing in-school professional development led by one or more school-based facilitators, focusing on building relationships with Maori students and using interactive classroom pedagogies such as cooperative learning. The professional development processes include classroom-based coaching and teacher-led inquiry groups that examine what teachers are doing to improve Maori student achievement.

Research on this subject is presently documenting a pattern of improvement in Maori students' achievement and well-being in school. Maori students of Te Kōtahitanga-trained teachers report much higher levels of satisfaction with and engagement in schooling than they had before the project (Bishop et al., 2006). Further, schools that have participated in Te Kōtahitanga for several years are posting considerably higher student achievement scores among both Maori and non-Maori students than are other comparable secondary schools in New Zealand (Maori in Mainstream, n.d.).

Te Kōtahitanga is not the only teacher professional development program that develops inclusive pedagogy through student voice, but it is perhaps the most extensive and researched project of this type. Cook-Sather (2006) discusses a preservice project—Teaching and Learning Together—based on the same basic philosophy. She points out that this approach to teacher education repositions those who occupy an institutional place as "least able and least powerful" into leaders and knowers, and those who occupy an institutional place as teachers, into learners. Teachers learn to listen, and students—especially those whom schools have silenced—learn to speak. In the case of Te Kōtahitanga, the students are explicitly conceptualized as culturally located beings; so, as teachers learn to listen to them and reshape pedagogy based on student voice, they also learn to include some knowledge from the wider Maori community in the classroom.

INCLUSION THROUGH COMMUNITY-BASED SERVICE LEARNING

In teacher preparation programs that include community-based learning (often organized as service learning), communities serve as co-teacher educators. Cross-cultural community-based learning means learning about a community that is culturally different from one's own by spending time there, equipped with learning strategies such as interviewing, active listening, and nonjudgmental observation. In service learning, the experience is designed specifically to serve community-identified needs. Marginalized communities serve as co-teacher educators when members help to plan the experiences and work substantively with teachers or teacher candidates. Research on the impact of cross-cultural community-based learning, although minimally found, underscores the potential of this kind of learning (e.g., Bondy and Davis, 2000; Brown, 2004; James and Haig-Brown, 2002; Melnick and Zeichner, 1996; Moule, 2004; Seidl and Friend, 2002; Wiggins et al., 2007). I will describe three such examples— two in the United States and one in Canada.

Ideally, community-based learning should be integral to the teacher preparation program as a whole; in reality, often it begins with an individual faculty member teaching a course. For example, working at the University of Indiana, Boyle-Baise (2002) developed a twenty-hour community-based service-learning component of a three-credit multicultural education course. The great majority of her teacher candidates were white; she wanted them to learn to collaborate with and learn from communities of color and low-income communities. Her work is particularly significant because of the collaborative power-sharing relationship that developed as a result. The community partners included: "two pastors, one for a racially mixed congregation, the other for a predominantly Black church; a director of a university program for students of color and education director for the black church; the program director for Boys and Girls clubs; the director of a community center; [and] the parent coordinator for Head Start" (Boyle-Baise, 2002, 78).

Boyle-Baise described her work as an ongoing process of building relationships, of "working with representatives of culturally diverse and low-income communities as coeducators for future teachers" (91). The community partners described the process of working with her as one of shared control, not only over what students did in the community, but also over the nature of the course, which they helped to teach. Coursework was structured around a community-based inquiry project, as well as in-class reflections over what students were learning. Each student worked out a written contract with his or her community partner; the partner helped to evaluate the student's performance for the course grade. Underlying these structures was time and effort invested in building ongoing communication and collaborative decision-making about the entire course and service-learning experience.

The Urban Teacher Education Center in Sacramento, California, involves a collaboration between the Teacher Education program at California State University, Sacramento, and the Sacramento City Unified School District. Created in 2004, this three-semester program is designed to prepare future educators for urban schools and communities. It is housed in an urban elementary school, where courses are held. Every classroom is assigned a pair of student teachers, and teacher education faculty members have become actively involved on school committees. By locating its work within an urban school, faculty members have forged a much closer connection between theory, research, and practice, which is normally the practice in teacher education.

In addition, student teachers become active in the local community. The purpose of community involvement is to help student teachers learn to draw on cultural strengths and resources of urban communities and families when teaching urban children in the classroom. During their first semester, they complete a community study "in which they get to know the community, the neighborhoods, and the public housing complexes in which the children and families live" (Noel, 2006). To do so, they gather information about the community by interviewing some key adults, meeting parents, visiting a local church, riding public transportation, and so forth. According to Noel (2006), the program's coordinator, the most significant community partnership has been with an after-school mentoring academy that is located in a housing complex near the school and was founded by two men who grew up there. Student teachers act as program tutors and mentors, enabling them to learn about the out-of-school lives of children in their classrooms, interpreted through adults who live in and grew up in the local urban community.

Noel (2008) points out that significant issues related to power and privilege must be confronted when collaborating with historically marginalized communities. Such communities have histories of unstable relationships with mainstream organizations that come and go according to their own needs, and of working with agendas set by others who presume to know what these communities need. To confront these issues and strengthen relationships with community members, Noel spent a sabbatical in the community. Like Boyle-Baise, she worked hard to develop the trust and communication that enabled the community to take ownership over a portion of the teacher education experience.

In Toronto, Canada, York University's Urban Diversity (UD) Teacher Education Program, which has been in operation since 1994, institutionalizes similar work, but on a larger scale. The UD Program is designed to prepare teachers through experiences that link schools, university, and urban communities. Community involvement is integrated throughout the entire program as a basis for learning culturally relevant practice and critical social analysis. As faculty members who are involved with the program explain,

> In the foundational dimension of the curriculum, candidates were introduced to the study of the social, cultural, political, and economic forces in the larger community that affect the pedagogical process; the concept of community-based teacher education with unique service as well as learning aspects; and theory-practice linkages, critical reflective practice, and an evaluation process that provides the structure for candidates to think, talk, and write. (Solomon, Khattar Manoukian, and Clarke, 2005, 175)

The program intentionally recruits teacher candidates from widely diverse racial, ethnic, linguistic, religious, and social class backgrounds. Learning to plan and share with each other prepares teacher candidates "to move across institutional borders: from the university to the practicum school to the community" (Solomon, Khattar Manoukian, and Clarke, 2007, 73).

As a part of the UD Program, candidates are required to participate in a community-based project, of which there are four types: health and safety, educational, recreational, and political. For example, health-oriented breakfast and snack programs give teacher candidates "the opportunity to interact socially and academically with students, teachers, and parents, while simultaneously becoming more aware of issues of poverty and social class and their role as border crossers" (Solomon, Khattar Manoukian, and Clarke, 2007, 74). A program that is classified as political is a women's shelter, where teacher candidates tutor children while simultaneously learning about domestic violence and poverty in the local community. Prior to their work in the community, the teacher education program prepares candidates with research skills that emphasize interviewing (and listening), observing, and document analysis, as well as learning to "bracket their assumptions about urban, inner-city communities" (75).

Based on an investigation of the impact of this program, Solomon, Khattar Manoukian, and Clarke (2005) found the nature and extent of teacher candidates' learning to vary widely. While some continued to see community involvement as extracurricular and maintained a charity-work stance toward it, others came to see the community as a valuable partner in education and community involvement as political work. Some candidates saw themselves as distant outsiders to the community, while others learned to navigate the structures of privilege and cultural identity to enter into substantive dialog with community members. Like Noel (2008), Solomon, Khattar Manoukian, and Clarke (2007) emphasize that issues of power and privilege, visible in this kind of program, must be confronted. The most significant tension is that universities and university students bring assumptions and privileges that often undercut inclusion and solidarity. The authors note, "It is often the case that initiatives are taken without direct negotiation with community members or social agencies that operate in the community. Such actions often alienate the very people they are supposed to serve. Moreover, they confirm suspicions that those outside the community who have little vested interest in the community can engage in actions that directly affect community members without having to be accountable to the community" (82).

As these three examples illustrate, teacher education programs can build collaborative relationships with communities in order to engage teacher candidates in cross-cultural community-based learning. Building such relationships requires ongoing communication in which community needs must have priority, and community members must have some say about the substance and process of teacher education.

Pedagogies of Inclusion and Political Consciousness-Raising

Under neoliberalism, not only are peoples relocating globally on a massive scale, but wealth also is undergoing marked redistribution. Liberal policies generally emphasized opportunity and competition, moderated by protections against discrimination and market excesses. Under neoliberalism, the role of government shifts from regulating markets to enabling them, and from providing public services to promoting private enterprise. The result has been a massive redistribution of wealth upward, or, as Harvey (2005) put it, a new restoration of elite power. Pedagogies of inclusion need to rest on an analysis of these rapid political and economic dynamics, enabling teachers not only to understand them but also to envision organized counter-action (Compton and Weiner, 2008).

Collaboration with historically underserved and immigrant communities has the potential to raise teacher candidates' political consciousness, but this may not happen unless political conscious-ness-raising is made an explicit part of the curriculum. An example from the United States Illustrates this experience. Duncan-Andrade and Morrell (2008) work simultaneously with youth in urban schools and communities, and also with teacher candidates. Their work is situated in "critical counter-cultural communities of practice," which they define as pedagogy intentionally designed to counter "the exis-tence of a dominant set of institutional norms and practices" (11). Duncan-Andrade and Morrell explain that urban pedagogy needs to recognize "the conditions of inequality and the desire to overturn those conditions for oneself and for all suffering communities as the starting point and motivator for the urban educator and for the urban student" (10). Rather than assuming that gaining a conventional education will help students from marginalized communities move into the mainstream, a critical countercultural community of practice begins by focusing directly on structural and material inequalities in the school and the larger community, engaging students in a cycle of praxis that involves researching a problem, and then formulating, implementing, and evaluating a plan of action to address it.

To prepare teachers for this kind of pedagogy, Duncan-Andrade and Morrell (2008) recommend not only that coursework and class-room experience focus on critical social theory and practice, but also that teacher candidates' access learning spaces outside schools, in which critical countercultural communities of practice already exist. Examples include "after-school dance and theatre programs, sports leagues, community-based organizations, or tutorials," in which adults in the community work with youth to address real community issues (183). Both these authors work in programs in which teachers collabo-rate with urban youth to research and act collectively on local struc-tural problems of inequity, projects that involve both research into the community and collaboration with the community. Coursework linked with community research helps teacher candidates to situate local problems within a larger analysis of power and to connect action addressing local problems with other existing organized action. In this way, teacher preparation is linked with community empowerment.

CONCLUSION

Pedagogies of inclusion must be situated within the context of rapidly growing diversity within schools and communities, histories of oppression that many communities have experienced, and impacts of the expansion of global capitalism under neoliberalism. This is a tall order for teacher education, particularly in light of cutbacks that many teacher educators are experiencing.

In this chapter, I have suggested three pillars on which teacher education for pedagogies of inclusion can be built: the university, the classroom, and the community. I have argued that communities, although absent from most teacher education programs, are critically important, particularly in light of cultural and political dynamics today. Immigrant and historically marginalized communities not only care deeply about the education of their children and house cultural and knowledge resources that teachers need, but also bear the brunt of deleterious effects of neoliberalism. As the Study Commission (1976) suggested over forty years ago, human welfare stands to benefit when local communities are strengthened; schools can be a part of that process. Examples in this chapter illustrate how teacher education, by bridging communities, classrooms, and university, can play a powerful role in strengthening teaching and forging collaborative relationships between teachers and the communities in which they work.

REFERENCES

Bishop, R. and Berryman, M. (2006). *Culture Speaks.* Wellington, Aotearoa New Zealand: Huia Publishers.

Bishop, R., Berryman, M., Tiakiwai, S., and Richardson, C. (2006). "Te Kōtahitanga: The Experiences of Year 9 and 10 Maori Students in Mainstream Classrooms." *Ministry of Education.* Retrieved May 29, 2007, from http://www.minedu.govt.nz/goto/tekotahitanga.

Bondy, E. and Davis, S. (2000). "The Caring of Strangers: Insights from a Field Experience in a Culturally Unfamiliar Community." *Action in Teacher Education,* 22(2), 54–66.

Boyle-Baise, M. (2002). *Multicultural Service Learning.* New York: Teachers College Press.

Brown, E. L. (2004). "What Precipitates Change in Cultural Diverse Awareness During a Multicultural Course?" *Journal of Teacher Education,* 55(4), 325–340.

Compton, M. and Weiner, L. (eds.) (2008). *The Global Assault on Teaching, Teachers, and Their Unions.* New York: Palgrave Macmillan.

Cook-Sather, A. (2006). "Change Based on What Students Say: Preparing Teachers for Paradoxical Model of Leadership." *International Journal of Leadership in Education*, 9(4), 345–358.

Darling-Hammond, L. and Bransford, J. (eds.) (2005). *Preparing Teachers for a Changing World*. San Francisco: Jossey-Bass.

Duncan-Andrade, J. M. R. and Morrell, E. (2008). *The Art of Critical Pedagogy*. New York: Peter Lang.

Ejbari, N. and Soriano, E. (2006). "Migración, interculturalidad y mediación." In E. Soriano Ayala, A. J. González Jiménez, and M. M. Osorio Méndez (Coords). *Conviviencia y mediación intercultural*. Almería, Spain: Editorial Universidad de Almería. 153–156.

Feiman-Nemser, S. and Buchmann, M. (1985). "Pitfalls of Experience in Teacher Preparation." *Teachers College Record*, 87(1), 53–65.

Feistritzer, E. (1999). *The Making of A Teacher: A Report on Teacher Preparation in the U.S.* National Center for Education Information. Retrieved September 18, 2006, from http://www.ncei.com/MakingTeacher-rpt.htm.

Harvey, D. (2005). *A Brief History of Neoliberalism*. New York: Oxford University Press.

James, C. E. and Haig-Brown (2002). "Returning the Dues. Community and the Personal in A University-School Partnership." *Urban Education*, 36(2), 226–255.

Katz, S. R. (2005). "Emerging from the Cocoon of Romani Pride: The First Graduates of the Gandhi Secondary School in Hungary." *Intercultural Education*, 16(3), 247–261.

Ladson-Billings, G. (2006). "From the Achievement Gap to the Education Debt: Understanding Achievement in U.S. Schools." *Educational Researcher*, 35(7), 3–12.

Lyall, K. C. and Sell, K. R. (2006). *The True Genius of America at Risk*. Westport, CT: Praeger.

Maori in Mainstream (n.d.). *Te Kotahitanga*. Wellington, NZ: Ministry of Education. Retrieved July 9, 2008, from http://tki.org.nz/r/maori_mainstream/tekotah_e.php.

Melnick, S. and Zeichner, K. (1996). "The Role of Community-Based Field Experiences in Preparing Teachers for Cultural Diversity." In K. Zeichner, S. Melnick, and M. L. Gomez (eds.) *Currents of Reform in Preservice Teacher Education*. New York: Teachers College Press. 176–196.

Meuleman, J. (2006). "Between Unity and Diversity: Construction of the Indonesian Nation." *European Journal of East Asia Studies*, 5(1), 45–69.

Moule, J. (2004). "Safe and Growing out of the Box: Immersion for Social Change." In J. Romo P. Bradfield, and, R. Serrano (eds.), *Working in the Margins: Becoming A Transformative Educator*. Upper Saddle River: Merrill Prentice Hall. 147–171.

Noel, J. (2006). "Integrating A New Urban Teacher Education Center into a School and Its Community." Online submission, *Journal of Urban Learning, Teaching, and Research*, 2, 197–205.

Noel, J. (2008). "A School-Based Urban Teacher Education Program That Enhances School-Community Connections." Paper presented at the International Roundtable on School, Family, and Community Partnerships. New York, 2008.

Openshaw, R. (1999). "Some Twentieth Century Issues for Twenty-First Century Teacher Educators." *New Zealand Journal of Educational Studies*, 34(2), 323–334.

Puiggrós, A. (1997). "World Bank Education Policy: Market Liberalism Meets Ideological Conservatism." *International Journal of Health Services*, 27(2), 217–226

Quilaqueo, D. (2006). Valores educativos mapuches para la formación de personas desde el discurso de *kimches*. *Estudios Pédagogicos*, 32(2), 73–86.

———. (2007). Saberes y conocimientos indígenas en la formación de profesores de educación intercultural. *Educar em Revista*, 29. Retrieved July 8, 2008, from http://www.scielo.br/scielo.php.

Rubenstein, G. (2007). "Building a Better Teacher: Confronting the Crisis in Teacher Training." *Edutopia*, 3(8). Retrieved July 22, 2008, from http://edutopia.org/magazine/nov07.

Seidl, B. and Friend, G. (2002). "Leaving Authority at the Door." *Teaching and Teacher Education*, 18(4), 421–433.

Solomon, R. P., Khattar Manoukian, R., and Clarke, J. (2005). "From an Ethic of Altruism to Possibilities of Transformation in Teacher Candidates' Community Involvement." In L. Pease-Alvarez and S. R. Schecter (eds), *Urban Teacher Education and Teaching*. Mahwah, NJ: Lawrence Erlbaum Associates. 171–190.

———. (2007). "Pre-Service Teachers as Border Crossers: Linking Urban Schools and Communities through Service Learning." In R. P. Solomon and D. N. Sekayi (eds), *Urban Teacher Education and Teaching*. Mahwah, NJ: Lawrence Erlbaum Associates. 67–68.

Soriano, E. (2008). "The Educational Capability of Immigrant Associations in a Multicultural Context." Paper submitted for publication.

Soriano, E. and Ejbari, N. (2006). "La práctica de la mediación intercultural." In E. Soriano Ayala, A. J. González Jiménez, and M. M. Osorio Méndez (Coords). *Conviviencia y mediación intercultural*. Almería, Spain: Editorial Universidad de Almería. 157–162.

Study Commission on Undergraduate Education and the Education of Teachers (1976). *Teacher Education in the United States: The Responsibility Gap*. Lincoln, NE: University of Nebraska Press.

Suárez-Orozco, M. M. (2001). "Globalization, Immigration, and Education: The Research Agenda." *Harvard Educational Review*, 71(3), 345–365.

Thorat S. K. (1999). "Poverty, Caste and Child Labour in India: The Plight of Dalits and Adivasi Children." In K. Voll (ed.), *Against Child Labour: Indian and International Dimensions and Strategies*. Delhi: Masai Book and Third Millennium Transparency.

Uhn, C. (2007). "Toward A Multicultural Society?" *Korea Journal* (Winter), 5–7.

Vidali, E. L. and Adams, L. D. (2006). The Challenges of Globalization: Changes in Education Policy and Practice in the Greek Context. *Childhood Education*, 82(6), 358–362.

Wiggins, R. Z., Follo, E. J., and Eberly, M. B. (2007). "The Impact of a Field Immersion Program on Pre-Service Teachers' Attitudes toward Teaching in Culturally Diverse Classrooms." *Teaching and Teacher Education*, 21, 653–663.

Yuen, C. Y. M. (2002). "Education for New Arrivals and Multicultural Teacher Education in Hong Kong." *New Horizons in Education*, 44/45, 12–21.

Zeichner, K. (1996). "Educating Teachers for Cultural Diversity." In K. Zeichner, S. Melnick, and M. L. Gomez (eds), *Currents of Reform in Preservice Teacher Education*. New York: Teachers College Press. 133–175.

CHAPTER 10

Toward a "Critical Emancipatory Mezcla Praxis": Xicana Participatory Action Research in Teacher Education

Lourdes Diaz Soto

The idea that the academy can provide a space for social action projects within teacher education is common, as noted by 56,043 citations in High Beam Research (http://www.highbeam.com/). Most of these pieces relate to how teachers and schools can interface themselves within classrooms with curriculum, instruction, and occasionally within a framework of a "transformative pedagogy." This chapter calls for an extension of social action projects in context-specific teacher education projects that include community within a framework of participatory action research. More specifically it shares the evolution of what I will term "critical emancipatory mezcla praxis" as well as accompanying examples of several hope-filled projects students and I have pursued over a span of eight years.

Paulo Freire (1970) taught us that education could be an integral part of the practice of freedom. Coupled with Freire's advice/consejos, my students and I have pursued additional layers of pedagogical complexities by inserting critical feminist lens to the dialogue. The diverse students I have taught have taken the idea of finding a space for social justice in the democratic sphere seriously. We have collaborated as educators interested in viewing critical pedagogy as a means of ameliorating the struggles of communities in an increasingly complex social theatre.

Searching for Emancipatory Praxis

The term praxis was used by Freire (1970) to denote a synthesis of theory and practice where each informs the other in what Freire (1985) describes as the "dynamic where theory and practice make and remake themselves." Anzaldua's (1990) work helps to inform us about mezcla (hybrid) not only in terms of ethnicity (mestiza consciousness) but how to relate it to theory. Anzaldua's advice is timely for educators interested in pursuing the pedagogical dance of praxis:

> Necesitamos teorías that will rewrite history
> using race, class, gender and ethnicity as categories
> of analysis, theories that cross borders, that
> blur boundaries-new kinds of theories
> with new theorizing methods.
>
> (Anzaldúa, 1990, 25)

The feminist inclusion in our work (work by my students and I) has led us to pursue a feminist participatory action paradigm (Fem-PAR), which has more recently evolved into what we refer to as a Xicana Participatory Action Research (X-PAR) (Soto et al., 2008). The latter integrates Anzaldua's (1990) mestiza consciousness with Sandoval's oppositional consciousness (2000) and Delgado Bernal's (2001) notion of "cultural intuition." The inspiration from Freire, Anzaldua, Sandoval, and Delgado Bernal has led us to view mezcla (hybrid) theoretical perspectives that are in keeping with the social context. X-PAR, for example, was extremely useful in Texas where the community we engaged with was Mexican American, Mexican, or Xicana/Chicano. The idea of including "cultural intuition" in our work meant that we could move from largely patriarchal ways of seeing the world to learning to listen to our inner voices, to trust our intuition, and to interpret research outside existing paradigms. We began to envision our dreams and to communicate our ideas.

The idea of mezcla (hybrid) can also be related to Kincheloe and McLaren's (2005) descriptions of bricoleurs as we carefully chose a mix of methodologies and analytical tools to make sense of our work. "In the active bricolage, we bring our understanding of the research context together with our previous experience with research methods" (317). This is a reflexive process that includes not only looking at ourselves in the roles of researchers and practitioners but also in an equitable dialogue with community.

Affording opportunities for the practice of freedom within the context of teacher education can lead to a pedagogy that is humanization while at the same time affording learners/partici-pants/coresearchers concientization. Our inspiration comes from our continued, consistent, persistent praxis helping us to realize that educational-social action projects can emerge from decolonizing perspectives that respect existing linguistic-cultural-socioeconomic traditions of our coresearchers (participants, learners) and commu-nity members.

Our interest in feminist participatory action research (Fem-PAR) is based on our work with Frierian readings and community proj-ects. The lessons learned from a Frierian pedagogy have inspired us to work toward action rather than passivity. This counter-hegemonic approach challenges existing research models. We are not interested in studying the world, but rather in transforming the world. Within the idea of action, we have pursued participatory action research and more recently feminist participatory research. The beauty of partici-patory action research is that it is dialogic and relies on community involvement. This paradigm insists on equitable collaboration expect-ing the "experts"' to engage as facilitators and transformers of exist-ing undemocratic conditions. Literature that teaches about feminist social action includes, but is not limited to, the work by Brydon-Miller, M., Maguire, P. and McIntyre (2004), Kaplan (1997), and Olesen (2003).

The evolution toward X-PAR came as a result of working with a remarkable group of graduate students in Texas. The students asked for coursework viewing the Chicana experience. I submitted two courses in which we were able to concentrate on learning from Chicana theorists. I refer to this experience as "standing on the wings of the Chicana theorists." Anzaldua's work was paramount to our explorations, as was Sandoval's.

My students and I have learned that this work requires deep lis-tening, deep respect, and deep ethics. Issues of power become para-mount to our discussions, and decisions ensuring social justice and equity remain uppermost in our minds.

GRADUATE STUDENTS EXPLORING POSSIBILITIES/PROJECTS

I have been impressed over the years by the significant ways students in my courses have involved themselves in a multitude of projects. Some projects have had a greater impact than others, but all have been designed by learners who have demonstrated a desire to pursue social justice and equity.

bell hooks (2003) reminds us that education is a hope-filled endeavor:

> My hope emerges from those places of struggle where I witness individuals positively transforming their lives and the world around them. Educating is always a vocation rooted in hopefulness. As teachers we believe that learning is possible, that nothing can keep an open mind from seeking after knowledge and finding a way to know. (xiv)

One project involved Southeast Asian women who were brought to the United States as domestic workers by privileged international ambassadors. We read in the local newspaper that an employer threw one of the women out the window as she was beginning to reveal the extreme dehumanization the workers were experiencing. One student in particular was touched by this incident and decided to pursue the matter further, and she visited the domestic workers in the nearby community (northeastern city). Alita (pseudonyms throughout) visited women in the neighborhood and asked what were the concerns and how might she be of help. Several students joined her, but basically this was Alita's project. The women shared their frustrations about not understanding their rights, not being able to speak the language, and showed examples of extreme abuse on the part of their employers. Alita shared with us the struggles of the women who lived under the tyranny of their bosses.

The list of hardships seemed endless and alarming, including, but not limited to, receiving harsh punishments, constant threats of deportation for minor transgressions, beatings, lack of health care, none of the promised monetary compensations, and mental abuse. The women asked Alita for ways they might learn English, receive legal advice, and a place to meet away from their employers. Alita and her teammates followed up on all of the requests by providing legal advice, ESL instruction, and a storefront for the women to meet. Alita's greatest pride was when the women themselves led a press conference in English and marched down a major street in the city, demonstrating their struggle to the public at large. What a victory this was for the women and, needless to say, for Alita. The struggle continues as Alita writes grants for the women and continues to support the group.

Another group of students in Alita's class was challenged by the concern for the multiple struggles faced by homeless children. This group of students met with the most resistance from shelter administrators as they attempted to alleviate the suffering of these children. Two students involved in this project visited shelter sites to obtain entry and interview the children affected by this problem. Their intent was to listen to the children's voices—the experts—the ones who were most affected by the situation. At that time there were estimates of 13,000 homeless children in this northeastern city. The shelter administrators viewed the students with suspicion, and the students reported that they seemed to be getting nowhere. Ultimately, this team decided to design an educational Web site that would provide not only data and statistics but also ways that average people could become involved and help the plight of homeless children. Understanding the sense of hopelessness with this issue led the pair to brainstorm about the possibilities and sharing their ideas with others. The Coalition for the Homeless estimated that there were 13,118 homeless children in New York City alone.

One of the more important issues that was discussed is housing. What does it cost to house a family? What is the ultimate cost to *not* house a family? What are the complexities of educational opportunities as well as health coverage? In a city in Texas, for example, the estimates are between 5,000 and 6,000 homeless children, at any given time. Are there ways teacher education projects can involve students in active participation leading to dialogic possibilities? Is this the world you have been dreaming about?

The treatment of children of color is of utmost concern for many of us. The Eastside School in an affluent city in Texas, for example, was built on a dumpsite. The graduate students in this class conducted a research, which indicated that methane gas may be impacting children's health and well-being. They wrote to the local Spanish language newspaper and immersed themselves in a variety of activities preparing themselves to share the information with the local community. A parent whose children were to attend this school in the near future was the one who raised this concern. The fact that a similar incident occurred on the more affluent west side of the town but was quickly remedied by building a new school led the students to conclude that this issue constitutes environmental racism. The school claimed that testing conducted by environmental experts shows that the gases being emitted are within the legal limits. They continue to test the school site, leading the students to back off from this project. As far as I am concerned, this project will continue to be in progress as long as children's well-being and health are at stake.

The visual and performance arts have also played an important role in several teacher education projects. In a northeastern city, for example, several projects were initiated that involved dancing and drama. A dance group for African American girls and a children's drama group with civic lessons are among the most unforgettable. The drama group is based upon spiritual principles, so that each time the children rehearse for a play they will also ponder human lessons. These experiences brought joy to the community as they could see their children gain confidence and a means of expressing themselves. In an increasingly testing-oriented curriculum, we seem to forget that there are tremendous possibilities in the arts for learning. Watching children sharing their gifts with pride serves to remind us of our ethical responsibility to ensure children's health and happiness.

In central Pennsylvania, graduate students took it upon themselves to design a three-dimensional collage with artifacts at the student union building that would serve as a teaching experience similar to an art exhibit. This project came about as we realized that an administrative assistant at this institution who died earlier was never mentioned nor recognized for any of her struggles or contributions to that university. Everyone carried on as usual in "Happy Valley." The students thought about the number of women's voices that are often disregarded and wanted a way to highlight women's role. This was their solution.

The collage was encased in a prominent location in the first floor of the student union building. It depicted multiple sociohistorical colonial issues of race, class, and gender. The topics I remember most clearly included the sterilization of Puerto Rican women, where data estimate that over one-third of the total population was sterilized during various campaigns. The students depicted the typical "batitas de casa" (little house coats) and inserted narratives about this horrific genocide. The socioeconomic struggle of Mexican mothers was depicted with family pictures and narratives demonstrating the historical mistreatment of Mexican-Americans, Xicana/Chicanos, and Mexicans in the United States. This struggle continues as immigrants are demonized rather than humanized. The anguish of the Korean comfort women during the Japanese occupation was also illustrated with artifacts and historical documentation. We had our photo taken in front of the glass case and honored all of these women.

It was the fall semester in Happy Valley when Anita (pseudonym) came to class extremely upset. She shared that she had not slept the previous night because she viewed a video that disturbed her. She asked if she could share it with the class. She brought the tape entitled "Senorita Extraviada" the following week. The video depicted the disappearance of hundreds of young Mexican women in the maquiladora district of Juarez, Mexico. The murder of these young women has been recognized since 1993, yet there has been no resolution. This matter continues to haunt us. The students in that particular class helped to get the word out by showing the video to other students in that university, notifying Amnesty International, and notifying the *New York Times.* I was also able to play a copy of the video at a Reconceptualizing Early Childhood Conference (an international group of early childhood educators) in Tempe, Arizona, where additional allies learned about the situation. The Reconceptualists plan to march at a later time with the parents of the women who have been killed. It is not yet clear when these killings, which still continue, will be resolved.

I continued to show the Lourdes Portillo video to my students for several semesters. When students in Texas viewed the video, I realized how insensitive I had been. Sometimes I feel as if I am on automatic pilot as I choose activities and experiences for my students, and in this case there was yet another large lesson for me to learn. The sociocultural context was totally different in the Texas scenario and garnered very sad and exasperating reactions since we were located so much closer to Mexico. I will never forget the reaction of two of the students, and another who actually lives in the Juarez area. She considers this experience as an integral part of her daily lived reality. She told us how the community tries to "not talk about it." She shared how helpless she felt about the continued killing of the young women of Juarez. Ultimately we talked more about her experience in my office. One day she announced to me that she has decided to conduct her doctoral dissertation in her community and view this issue as an integral part of her research. This is not a solution to the problem faced by the young women of Juarez, but it has brought her a sense of strength about herself and her future possibilities.

Our discussions and assignments have also included performative pieces based on the work of Garoian (1999). The first time I tried to integrate performative elements into my teaching, I asked Charles to visit the class and share his vision (at Penn State). The students in that particular class (international adult educators) never quite understood the intent of performing pedagogy. I subsequently borrowed and modified one of the exercises in Garoian's book (220) in two seminars that I taught with both graduate and undergraduate students. The exercise is a "live collage" where each of the students prepare for a brief performance involving their personal memories for an autobiographical presentation. There are several elements involved in this, including performing a task that is very familiar yet unrelated to the narrative, projecting images also unrelated to the narrative, plus the verbal story component.

In the past, I used this activity at a later time in the semester. Recently I have found that it is more powerful as a way of introducing the group to each other and beginning to create community within the classroom space. I ask students to ensure confidentiality and to share only what they feel comfortable sharing. Students' willingness to share their lives with their peers has taught us about their complex struggles within the academy. I remember Nadia reflecting on what it is like to live in a car. How she hid her situation as if it were a crime and continually felt "less than" all of her peers. Joel told us about how his younger brother was killed by the local gangs and how he promised himself that education was the better choice for him. Nilda relayed her aunt's struggle as a lesbian living in a rural community where she was disrespected and ridiculed. These are just a few examples to demonstrate that the humanity of our students needs to be uppermost in our minds in teacher education. Providing a space where students feel comfortable with each other as allies (as opposed to being combatants) is crucial in order to open a compassionate dialogue. We could observe through windows into the lives of our presenters. I have found this activity quite remarkable.

The last two years working with eastside families in a Texas city has also been illuminating with regard to the lives of immigrant children and families (Soto et al., 2008). My students and I have been reminded of how often immigrants are dehumanized, especially when we witnessed children incarcerated in the Hutto Detention Center in Texas merely for being immigrants. We collected children's illustrations and learned that the children have a tremendous amount of information about immigration. Children are aware of border crossing politics, they are cognizant of the heroic efforts made by people to cross the border, they are conscious of people being deported, they are able to depict issues of power and authority as related to immigration officials, they question authority, they question elements of historical freedom, and they live a hope-filled existence of sharing power as shown in the immigrant marches. We also learned that children have been informed about the sociocultural and historical elements of colonization. According to Maribel (age eleven) "it was Christopher Columbus who changed the story." According to Valerie Polakow (2008), we were able to make visible "children's in vivo perspectives...the sense of injustice that frames their lives." We have begun to share this information at national and international conferences to highlight the experiences of immigrant children.

Students from this same institution have also designed a community platica (dialogue) attempting to bring together numerous organizations with similar goals. The breakout sessions from this experience showed the complexity faced by the community as well as the continued, persistent work for social justice and equity by young people.

My students have shared their passions and their visions because of the safety they have felt in the community of our classroom space. I have felt a particular "chemistry" with certain groups that have led to a conscientication, and I have asked that all of our conversations and dialogue be kept confidential as if we were a "family." Over the years, several groups have continued to feel this sense of being part of the "academic family" and have vowed to stay in touch and have regularly met with each other long after their classes ended. I have been honored to meet such incredible human beings committed to social justice and equity. They have become my inspiration and source of Freirian love. Some students have been "adopted" because they have sometimes come into my life and have almost become a close colleague or an ally. I have often felt that my students come second only to my biological family. They have brought me tremendous joy and pride. I thank each and everyone for sharing their critical feminist Freirian spirit that has led us to "critical emancipatory mezcla praxis."

I would like to suggest that a critical Freirian pedagogy coupled with elements of a feminist participatory research opens up a space of possibility for context-specific praxis where dialogue, democratic participation, creativity, and action/reflection/action can take place. Freire's (1985) notion of "reading the word and the world" is perhaps the most valuable element of a pedagogy that is critically Freirian and feminist. It is our intent to pursue a "critical emancipatory mezcla praxis" that, borrowing from Gandhi, will guide us "to be the change we would like to see in the world."

References

Anzaldúa, G. (1987, 1999). *Borderlands, la frontera: The new Mestiza*. San Francisco, CA: Aunt Lute Books.

Brydon-Miller, M., Maguire, P., and McIntyre, A. (eds.) (2004). *Traveling Companions: Feminism, Teaching and Action Research*. Westport, CT: Praeger.

CI Austin. Retrieved on August 10, 2007 from www.ci.austin.tx.us/health/ms_homeless_tf.htm.

Coalition for the Homeless. Retrieved August 10, 2007 from www.coalition forthehomeless.org/advocacy/basic_facts.html.

CWLU Herstory. Retrieved August 10, 2007 from www.cwluherstory.com/CWLUArchive/puertorico.html.

Delgado Bernal, D. (2001). "Using a Chicana Feminist Epistemology in Educational Research." *Harvard Educational Review*, 68(4), 555–582.

Denzin, N. and Lincoln, Y. (eds.) (2003). *The Landscape of Qualitative Research: Theories and Issues*. London: Sage Publications. 332–397.

Freire, P. (1970). *Pedagogy of the Oppressed*. New York: Herder and Herder.

———. (1985). *The Politics of Education*. South Hadley, MA: Bergin and Garvey.

Garoian, C. (1999). *Performing Pedagogy*. Albany: State University of New York Press.

hooks, bell (2003). *Teaching Community. A Pedagogy of Hope*. New York: Routledge.

Kaplan, T. (1997). *Crazy for Democracy: Women in Grassroots Movements*. New York: Routledge.

Kincheloe, J. and McLaren, P. (2005). "Rethinking Critical Theory and Qualitative Research." In N. Denzin and Y. Lincoln (eds.), *The Sage Handbook of Qualitative Research* (3rd ed.). London: Sage. 303–342.

Olesen, V. (2003). "Feminisms and Qualitative Research at and into the Millennium." In Perales, E. (ed.), *Mapping Possibilities: Xicana Participatory Action Research* (X-PAR). American Educational Research Association, New York City.

Polakow, V. (2008, April). Discussant remarks at the annual meeting of the American Educational Research Association, New York City.

Sandoval, C. (2000). "U.S. Third World Feminism: Differential Social Movement." In *Methodology of the Oppressed*. Minneapolis, MN: The University of Minnesota Press. 40.1–63.4.

Senorita Extraviada. Retrieved August 11, 2007 from www.lourdesportillo.com/senoritaextraviada.

Smith, L. T. (2005). *Decolonizing Methodologies: Research and Indigenous Peoples*. London: Zed books.

Soto, L. D., Cervantes, C. N., Milk, C., Campos, E., Garza, M., Godinez, D., et al. (2008, April). *Border Crossing Children's Perceptions of Immigration*. Paper presented at the annual meeting of the American Educational Research Association, New York City.

Stewards. Retrieved August 11, 2007 from www.stedwards.edu/newc/capstone/sp1999/homeless/Papers/Intro.htm.

CHAPTER 11

Equity Issues in Early Childhood Teacher Learning in Australia

Glenda Mac Naughton

INTRODUCTION

Pedagogies of exclusion have a long history that touches different nation states in different ways. In exploring equity issues in early childhood teacher learning in contemporary Australia, I would like to begin with a moment from the past beyond the borders of our nation state. It is a moment from the 1960s when U.S. black civil rights leader Dr. Martin Luther King, Jr. once said:

> In the End, we will remember not the words of our enemies, but the silence of our friends.

In June 2007, Australian troops occupied sixty indigenous communities in the Northern Territory at the behest of the then Australian Prime Minister John Howard. In response, Jennifer Martiniello, an award-winning writer, poet, and artist of Arrernte, Chinese, and Anglo-Celtic descent, wrote an open letter to major Australian newspapers. In her concluding paragraph, she wrote:

> As reported in the Sydney Morning Herald 25th June, the Howard Government last week used the military to seize control of 60 Aboriginal communities in the Northern Territory, which are now under military occupation. This is not Israel and Palestine. The Northern Territory is not Gaza or the West Bank. This is Australia—but is it the Australia you thought you lived in? Walk in our shoes, Aboriginal Australia's,

and ask yourselves, what would it be like to have this done to us? And then, walk with us. (http://www.publicenemy.com/pb/viewtopic. php?p=205191&sid=46457d5ea377b81f5515f3432e13b1f9) (Accessed June 19, 2007.)

The call to "walk together" for justice and human rights rings loudly through time and across diverse struggles against injustice and discrimination. It is a call by the colonized, the dispossessed, the oppressed, the marginalized, and the silenced; and it is a call for solidarity with them in their fight against injustice and toward justice. It is a call to disrupt unjust power relations and to learn to use our power for justice. It is a call that in the field of early childhood challenges us to challenge pedagogies of exclusion.

To respond to that call and create Australian early childhood as a space where we can "walk together for justice" and build pedagogies of inclusion, we need early childhood teachers who can learn to be critical meaning-makers. Critical meaning-makers refuse to naturalize existing unjust relations of power (McLaren, 1995) and work to change them. Critical meaning-makers remake pedagogies of exclusion into pedagogies of inclusion. They act as change agents in the reformation of pedagogies for equity.

To do this, critical meaning-makers must become intimate with power, understanding its discourses and its dynamics—how they work and whose interests are served by them. Discourses are the ideas, words, images, and feelings that shape how we make sense of the world, what we value in the world, and how we act in it. They shape what we believe is just or unjust and they shape how we exercise power in the world. Critical meaning-makers need to be intimate with the ideas, words, images, and feelings (the discourses) that work for and shape equity and respect for diversity, shape our desires to work for justice, and thus shape our desire to create pedagogies for inclusion. Three strategies help critical meaning-makers do this. Those strategies are:

1. Remembering the past...in order to disrupt its injustices in the present.
2. Revealing the discourses of privilege and the discrimination they breed...so as to reposition them.
3. Rethinking what I know...in order to transform what I/we do.

I want to show what these strategies mean in practice and how the early childhood teachers can use them in their work for equity and respect for diversity and build pedagogies of inclusion by returning to my beginning—the continuing military occupation of indigenous land in the Northern Territory of Australia.

Remembering the Past . . . in Order to Disrupt Its Injustices in the Present

The occupation of indigenous land in Australia at present has been made possible by our colonial past. As a critical meaning-maker, it is the past that I need to remember in order to disrupt injustices against indigenous Australians in the present. This is because it is in that past that governments first exercised their power to dispossess and oppress indigenous Australians. It is in the past that the racialized ideas, words, images, and feelings that made this racial injustice possible began (Atkinson, 2002; HEROC, 1997).

As I share my remembering to disrupt racial injustices in contemporary Australia, I invite you to reflect on the pasts that early childhood teachers may need to *remember to disrupt* injustices in your present to build pedagogies of inclusion. In this chapter, I argue that learning to remember and to disrupt is essential if early childhood teachers are to take a meaningful stance against inequities in contemporary Australian civil society that touch the lives of the children with whom they work and that inure in pedagogies of exclusion.

My remembering has two foci:
remembering "race" in *the* world
remembering "race" in *my* world.

My disrupting has two foci:
mapping the past in *our* early childhood present
mapping the choices this brings us.

Remembering the Colonial Past . . . in Order to Disrupt Racial Injustices in the Present

Remembering "Race" in the World

Ghandi (1998) defined colonialism succinctly as the "historical process whereby the 'West' attempts systematically to cancel or negate the cultural difference and value of the 'non-West' (16), relying on distinctions between the 'civilized' and the 'non-civilized.'" European colonists saw the original inhabitants of European colonies as non-civilized and "genetically pre-determined to inferiority" (Ashcroft, Griffiths, and Tiffin, 1998, 47).

Indeed, the British invented the term "aborigine" in the mid-1660s to distinguish indigenous inhabitants of its colonies from their "colonists." The point of those distinctions was to racialize indigenous Australians as the primitive savage who was inferior and yet to be feared. Similar processes of racialization were, and still are, at work in other colonized countries, with similar effects—colonizers always seemed superior to the colonized. It is a process that began over 400 years ago, but its effects and processes remain with us even today. As one Australian early childhood teacher-researcher explained:

Well, the Philippines is my country of origin and the Philippines suffered from dual colonisations. What happened was that as a result of these colonisations we have lost partly our language because English is used more now, it's more prevalent. It is our medium of instruction, it is the language of business conversations and it is the language of our textbooks. It is not only about loosing our identity and cultural roots it is also about taking in values and it's about how people think. For example, from the Spaniards we learnt about colorism—white is beautiful, white is intelligent, white is powerful. And when the Americans came to colonise they thought that it was important for us to [be] educated and to copy their education. The sad thing for us [is] that we have taken that as a gift. It is just a natural course of things and in a way we began to act as little brown Americans. Being born in a colonised family means that you inherit these values—it's like you breathe it, it's in the air, and it's normal for me to think this way—very western and having all these values. (Cultural Diversity Resource 2006, Video Transcript)

Remembering "Race" in My World

In the 1960s, while the black civil rights movement in the United States was at its height and while Australia's indigenous peoples struggled for civil rights and for land rights, I learnt about Aboriginal Australia through my primary school text—*The World We Share*—that "the aborigines…lived very much like the Stone Age people of other lands" and that "when white people first saw the aborigines, they thought them savages who lived more like animals than humans" (Pownall, 1960, 31). They appeared in the past tense, as primitive and as a strangely different "other." My textbook linked me effortlessly and seamlessly to colonial Australia and its European beginnings—the time and space in which indigenous people's human rights were denied and their land rights were (literally) nullified. That textbook both recalled the colonial era *and* reinstated it. It taught me about indigenous Australia by inviting me to accept uncritically the images and language of its colonization and to learn the discourse of colonialism. What do you remember of "race" in your world?

Mapping the Past in Our Early Childhood Present

In the ensuing forty years, much has changed, and yet so much remains unchanged. Many young white Anglo-Australian children continue to learn—as I did—to reproduce colonial distinctions between indigenous and nonindigenous Australians and between themselves and "non-white" others. Those colonial distinctions created by colonizing European nations over 400 years ago in the mid-1660s have survived into my Australian childhood and continue today in Australia. In the late 1990s, as part of the Preschool Equity and Social Diversity Project, my colleagues and I talked with young Australian children about what it means to be an Aboriginal Australian. As I have reported elsewhere (Mac Naughton 2005; Davis and Mac Naughton, 2001; Mac Naughton, 2001a, b, c, d), when the children described what it means to be an Aboriginal Australian, 71 percent of them did so from within the colonial discourses of race that I had learnt in the 1960s and that had begun over 400 years ago. To them, Aboriginal Australia was in the past, was strangely different, and, for some, was the fearful "other." Here are some of their responses:

Aboriginal Australia as "in the Past"
For example:

- Aboriginal people didn't usually wear clothes. Why is that one (a doll) wearing them?

Aboriginal People as Strangely Different
Because of their housing, their clothes, their food (and the fact that they caught it), their skin color, and their language.
For example:

- They have leaves for undies and no other clothes. They eat grubs.
- They live a long way away from me and talk another language.
- They don't eat the food that we eat.

Aboriginal People as the Fearful "Other"
For example:

> *Researcher*: What can you tell me about Aboriginal people?
> *Child*: They are bad. They will come and kill you in the night with their knives. They will kill you dead.

For these reasons, a key equity issue for early childhood teachers deconstructing pedagogies of exclusion and reconstructing pedagogies of inclusion in Australia is to ask the question, how does "race" in our past enter early childhood in our present? This is not without its challenges. In a small-scale exploratory study (see Mac Naughton and Hughes, 2007), we found that many early childhood teachers were uncertain about how best to respond to cultural and "racial" diversity in early childhood spaces. Alongside this there was a clear mismatch between social expectations that teachers would encourage children to respect that diversity and teachers' pedagogical practices.

Mapping the Choices This Brings Us

The fact that the Australian racialized colonial discourses exist in the space of public dialogue provided by Australian early childhood institutions implies a choice: do early childhood teachers learn to challenge such colonial discourses that have been used in this country and elsewhere to justify indigenous dispossession and oppression or do we remain silent and allow them to continue? To borrow from Martin Luther King: how should white Australia be judged for our silences and how should early childhood pedagogies be judged for their silences?

To walk with others for racial justice in Australia, early childhood teachers must learn to speak forcefully and persistently about the colonial texts that continue to reproduce colonial injustice—challenging what is present in these texts and highlighting what is silenced in them. In inclusionary pedagogies, they must learn to acclaim the reality of contemporary indigenous Australia in all its historical, material, cultural, geographical, spiritual, and linguistic differences. They must also learn to highlight how these long-term contentious relationships still disadvantage indigenous Australians. The current Australian Prime Minister's formal Apology to the Stolen Generations (February 13, 2008) created an important discursive shift in Australia's civil landscape. It formally recognized and apologized for disadvantages suffered by indigenous people now and recognized their links to a past where impacts and legacy of colonization include the loss of lands, health, family, and kin, and the destruction of indigenous languages, culture, and traditional lore.

We know that children can learn, with support from adults, to do the same, and in doing so they can learn to move beyond reinstating the colonial. For example: When Jay (four and a half years old) was asked about what he knew about being Aboriginal, he answered that he knew, "knew quite a lot actually":

> The Prime Minister (John Howard) doesn't like them. There's two halves, you see. On one half is the Aboriginals and on the other half is the Prime Minister. The Aboriginal half is the good half and the Prime Minister's half is the bad half because he wants to take their land away from them, you see.

Several people feel discord when I report these comments and have asked me:

> "Why were you talking to children about race at all?"
> "They are too young to know this stuff."
> "Their parents shouldn't be brainwashing them in this way."
> "That's not our business in early childhood—that's politics."

The debate about whether early childhood teachers should discuss racial politics with young children has engendered over seventy years of research about children and race, and it has established beyond a reasonable doubt that from as young as three years of age children know about "race"—their own race and that of others (e.g., Aboud and Doyle, 1996; Connelly, 1998; Hirschfeld, 1995; Johnson, 1992; Mac Naughton, 1993; Mac Naughton, 2001a, b, c, d; Ramsey, 1991). It is time to ask instead, "What do we want young children to know about race?" and, "What do we reasonably want them to say about it to themselves and to others?" if we are to create early childhood institutions as spaces of public dialogue where we can walk together for racial justice. How we answer these questions may differ across nation states, yet in Australia, against the formal Apology to the Stolen Generations, the continuing occupation of indigenous lands, and the continuing disadvantages faced by indigenous Australians, their relevance to equity and respect for diversity make them essential for Australian early childhood teachers to learn to ask, what do I reasonably want children to know about "race" in my context? Without tackling this question, pedagogies for racial equity cannot grow. More specifically, Australian early childhood teachers need to ask, what do I reasonably want children—indigenous and nonindigenous—to know about each other? Without solving this question, pedagogies that reform and reinstate indigenous exclusion will continue to grow.

Reveal the Privilege of Whiteness and the Racism It Breeds... So as to Reposition It

In learning to become critical meaning-makers who can build a contemporarily relevant and inclusionary pedagogy, early childhood teachers must learn to "remember the past...in order to recognise and disrupt its injustices in the present." As they learn to do this they must also learn to "reveal the discourses of privilege and the discrimination it breeds." So, as early childhood teachers learn to seek to disrupt racial injustices against indigenous Australians in our present by remembering our past and its connections to the present, they must also learn to reveal and reposition the discourses of privilege that hold such racial injustice in place. To carry out such revealing and repositioning, they must learn, in themselves and in their pedagogies, to:

- map the discourses of "race" that privilege racism in *the* world, in *their* world, and in the early childhood world,
- reposition those discourses of "race" to challenge racism.

Mapping Discourses of "Race" That Privilege Racism in the World, in My world, and in the Early Childhood World

Ien Ang identifies herself as an Australian woman of Chinese descent. Ang writes:

> It is important to emphasize, at this point, that white/Western hegemony is not a random psychological aberration but the systematic consequences of a global historical development over the last 500 years, and, it is this historical sense that the hierarchical binary divide between white/non-white and Western/non-Western should be taken account of as a master-grid framing the potentialities of, and settings limited to, all subjectivities and all struggles. (2003, 199)

Whiteness has been increasingly recognized as one of the most powerful discourses of privilege that holds racial injustice in place.

The task of an early childhood teacher as a critical meaning-maker seeking to create early childhood as a space for racial justice, therefore, is to find pedagogical pathways that "Reveal the privilege of whiteness and the racism it breeds...so as to reposition it" by mapping how it works to promote injustice.

Whiteness was implicated in colonialism and continues to be implicated in postcolonial societies such as in Australia and throughout Europe where the colonial was born. Recent research conducted by me and my colleagues in the Centre for Equity and Innovation in Early Childhood (the University of Melbourne) on children and "race" shows that whiteness is also implicated in what children learn is reasonable to think and to say—or silence—about themselves and others and what, in consequence, is reasonable behavior toward themselves and others, and therefore it is implicated in whether and how young children—and adults—learn to "walk together" for justice (Davis, Smith, and Mac Naughton, 2008; Mac Naughton and Davis, forthcoming).

Postcolonial scholars revisit, remember, and interrogate our colonial past (Ghandi, 1998) to trouble whiteness in order to reposition it. To explain this, I'll return to the early 1960s. At that time, while black rights were being fought for in Australia and the United States, my world was unremarkably white. My school texts influenced my view that whiteness was ordinarily connecting me with colonial views of indigenous Australians as the "primitive aborigine" (Pownall, 1960). All that was good and magical in my early years—Father Xmas, God, the Tooth Fairy, the Easter Rabbit, and my storybook princesses were all white. Nonwhiteness—the "other" to whiteness—appeared in the form of the exotic and strange "red Indians" and the silly and scary "black" golliwogs. I can't remember linking being white with being powerful, but the links between whiteness and power are indisputable, as Gussein Hage reminds us in his 1998 book, *White Nation*:

> No matter how much it is maintained that multiculturalism reflects the "reality" of Australia, the visible and public side of power remains essentially Anglo-White: politicians are mainly Anglo-White, customs officers, diplomats, police officers and judges are largely Anglo-White. At the same time, Australian myth-makes and icons, old and new are largely Anglo-White...Anglo-ness remains the most valued of all cultural capitals in the field of Whiteness. (Hage, 1998, 190–191)

Current CEIEC research is showing how many young children in contemporary Australia still see whiteness as good, desirable, and normal in two ways that repeat the patterns of my childhood. First, they regard being white as desirable—it was lovely and pretty, whereas dark skin "looks dirty" or "yuk." Second, the children talked about "real" Australians being white and about how skin color mattered to being Australian. For example:

> *Shana*: . . . She's (Olivia's) got the lightest. Do you know why she's got the lightest?
> *Researcher*: Why?
> *Shana*: Because, I bet she was born in Australian people. I already know that they are Australian, 'cos they have white skin. *They*'ve got different skin to Australian people.

> *Kylie*: What's the same as you and Olivia?
> *Fairy 1*: She has the skin same, skin like me too?
> *Kylie*: She has the same skin like you? What colour is her skin?
> *Fairy 1*: Normal.

Fairy 1 and Shana were four years old at the time of those conversations, but they had learnt already how to position themselves within the black-white binary that expresses colonial race-color. As Ravenscroft (2003) reminds us:

> Whiteness of course does not lie in the colour of one's skin but is a structural position, a position of privilege, the dominant term in a white-black dualism.

We found that 40 to 50 per cent of the Anglo-Australian children in our research considered white skin as "normal," "lovely," "best," and "Australian." What are the discourses of racial privilege that exist in your early childhood worlds?

REPOSITIONING WHITENESS IN EARLY CHILDHOOD

I suggested earlier that to create early childhood as a space of public dialogue where we can walk together for racial justice we need pedagogies that reveal the privilege of whiteness and the racism it breeds, so as to reposition it. We can build such pedagogies by talking about whiteness amongst ourselves and with children—discovering what children know about whiteness and what they believe can be reasonably revealed or should be concealed about whiteness, engaging them with questions of what is fair and not fair, and with questions of participation and nonparticipation, inclusion and exclusion, representation, misrepresentation, and nonrepresentation. Early childhood teachers can help children to become critical meaning-makers who understand, as Kiana does, "that people think them's the rules, but they're not."

> *Adult*—When we see the movies the baddies are mostly always in black.
>
> *Child*—Yes.
>
> *Adult*—Why do you think that they do that?
>
> *Child*—They think that that's the rules, the rules that are in the story but it isn't…because people, because baddies can have different colors.
>
> *Adult*—Why do the people in the movies always make people with dark skin to be in dark places and to be baddies?
>
> *Child*—Cos I think that they think that's the rules but it isn't.
> (Cultural Diversity Resource 2006, Video transcript)

A key question that early childhood teachers learning to become critical meaning-makers who build inclusionary pedagogies should ask and to learn to answer is, "How is whiteness positioned in your early childhood world?" To be a critical meaning-maker who builds inclusionary pedagogies for racial justice, Australian early childhood teachers need to learn to pedagogically reveal and reposition whiteness to rethink what they/we know, so they can transform what they/we do.

Rethink What They/We Know… in Order to Transform What I/We Do

For the early childhood teachers to learn to become critical meaning-makers, they need to learn to rethink what they know in order to transform what they/we do. They not only need to remember, reveal, and reposition discourses of privilege but also need to actively choose to privilege what I call "otherwise" meanings for justice and use these to guide pedagogical decisions. Otherwise-meanings are those held by the colonized, the dispossessed, the oppressed, the marginalized, and the silenced about the discourses and dynamics of power and their workings. Their meanings produce an "other wisdom" about justice and injustice and what it will take for us to walk together for justice.

To show you what seeking the otherwise looks like in practice, I want to tell you a story—called Kim's Blushes. The story is of four-year-old Kim, a Vietnamese-Australian child in a CEIEC project. As the story starts, I am sitting in the room ready to take notes, and Heather brings Kim into the room to meet our research dolls for the first time. Kim held Heather's hand tightly. She focused her attention quickly on the dolls. I was not sure which dolls held her attention so tightly, but I could see her staring at them. Kim sat down and listened closely as Heather introduced her to each of the dolls. As Heather talked about each doll, Kim looked at it closely. Heather then told Kim that we'd like to ask her some questions.

Kim: (blushes)
Heather: Do you understand, Kim?
Kim: [NODS]
Heather: When you look at the dolls, can you tell me which doll you think looks most like you?
Kim: [SILENCE. SHE LOOKS AT HEATHER THEN LOWERS HER EYES AND POINTS AT WHITE-SKINNED, BLOND OLIVIA. AS SHE DOES SO, SHE BLUSHES VERY STRONGLY.]
Heather: I see. Can you take a good look for me and be sure?
Kim: [NODS, THEN POINTS AGAIN AT OLIVIA, THIS TIME HOLDING HEATHER'S GAZE. SHE BLUSHES AGAIN.]

Why did Kim make the choice she did and why did she blush in making this choice? Was her response a desire for "whiteness" or not?

I have told the story of Kim's Blushes to diverse audiences in Australia, the United States, the UK, New Zealand, and Singapore. My questions—"Why did Kim make the choice she did and why did she blush in making this choice?"—have evoked various responses. I have been asked:

- Have you considered that it's impossible to answer the questions without more information about Kim, her background, her language skills, and so on?
- Could you have framed the research questions in ways that confused or were misleading?
- Did Kim think that that is how she looked because she doesn't see skin color as an issue? Is it possible that *you* are the ones making it an issue?
- Did Kim just want to go to the toilet or go out and play?
- Aren't you making a lot out of something that wasn't really that important?

Each response came from colleagues who are "white" Australians as I am, or are "white" North Americans, New Zealanders, and

Britons. But, I also have another set of comments that I want to share with you:

- I could feel my skin creep as you told Kim's story—I just knew what she felt. (indigenous Hawaiian female)
- I felt I wanted to cry because I knew those blushes well. (Japanese-American female)
- Very powerful. I am truly emotionally feeling a bit teary eyed and choked up when I read this. (Indian-American female)
- I know how Kim felt. When I was four years old I asked my mother to wash my skin so that the colour would come off. I was the only dark skinned child in my kindergarten. I want to thank you for telling Kim's story. It is important that people know how it feels. (Indian-Australian female)
- You know—that story has meaning for me. My niece, she is now six years old. When she was four years old she came home from kindergarten and at bath time she asked her mother to wash her skin that it would be lighter. (Singaporean-Indian female)
- Kim's story is very true for me. I always wanted white skin. (Malayian-Australian female)
- You have just told the story of my sisters. My wife and I have been close to tears as you talked but you have said what needed to be said (indigenous Australian man). (There followed a long discussion about how his sisters had spent their lives trying to pass as white, and the tragic implications of their sense of not being able to "pass," which had included drug abuse and suicide.)
- It hits you where you live. I felt very tearful when I read it. (indigenous Australian female early childhood consultant)

To rethink what we know we must search beyond what we know, asking who might know "otherwise" to me. Clearly, those who know "race" otherwise to me as an Anglo-Australian are those who directly experience its injustices. Through Kim and other children and adults, I have learnt that whiteness matters in the lives of each of us; that it matters differently depending on our relationship to it and its injustices; and that this gives us different insights into what knowledge is valid and whose knowledge needs to enter the space of public dialogue in early childhood institutions. Those who experience racial injustices—people of color across national borders—point to the intimate connections between skin color, experience, desire, and emotion in their responses to "Kim's Blushes." Explanations *other* than skin color link white, Anglos like me across national borders. To privilege "otherwise" is to see these nonwhite, non-Anglo, non-Western, non-colonial understandings of "race" and emotion as valid and valu-able guides to what is needed to walk together for racial justice as early childhood professionals. If I choose to privilege the "otherwise" in understanding "Kim's Blushes," I am drawn to know that "race" matters to young children, I am drawn to ask, "What must we do in early childhood to make 'racial justice' matter?" and I am drawn to question, "How can we create racially just white identities in our work with children?"

These are challenging questions, but to ignore them is to make racism and the whiteness upon which it sits live on in our lives and in the lives of young children who enter that public space of dialogue that is early childhood. There are many guides to follow in seek-ing to answer and act amongst these or other questions we might generate with early childhood teachers. Each relies broadly on the principles and practices of antidiscriminatory education (also known as antibias education and Respect for Diversity education). These approaches engage teachers and children in the study of power in the daily relationships with each other and in the practices of critical meaning-making together, so they can become activists for social justice in their daily lives with each other and in their wider community.

FINAL REFLECTIONS: A TRIAL AND AN INVITATION

Activism for social justice arises for many reasons, and the choice to actively seek social justice and equity in and through pedagogical practices is both deeply personal and politically inspired. Central to activism within education has been the process of conscientization that Freire (1970) referred to as "developing consciousness, but con-sciousness that is understood to have the power to transform real-ity" (Taylor, 1993, 52). Concientization directs a critically reflective stance on educational work toward action for social justice and equity. Pedagogical critical meaning-making seeks to do this. Early child-hood *can* be a space of public dialogue where we find ways to walk together for justice if early childhood teachers can learn to engage in concientization. I believe they can do this by becoming competent critical meaning-makers who forcefully and persistently in their peda-gogical practices find pathways to:

1. Remember the past in order to disrupt its injustices in the present;

2. Reveal the discourses of privilege—of race, gender, sexuality, class, geography—and the discrimination they breed, so as to reposition them;
3. Rethink what we know in order to transform what we do.

Becoming a critical meaning-maker building inclusionary pedagogies amongst civil landscapes of exclusionary politics may not be safe, politically or in a popular manner; but let me return to Martin Luther King:

> Cowardice asks the question, "Is it safe?" Expediency asks the question, "Is it politic?" Vanity asks the question, "Is it popular?" But conscience asks the question, "Is it right?" And there comes a time when one must take a position that is neither safe, nor politic, nor popular, but one must take it because one's conscience tells one that it is right. (Speech, "Remaining Awake Through a Great Revolution") (March 31, 1968)

I invite you to ask yourself—what might it take for the early childhood teachers in your contexts to learn to build pedagogies to "walk together" against injustice and toward justice? What critical meaning-making do they need to do now? What do you need to remember, reveal, and rethink for justice? How are pedagogies to be reconstructed to allow space for these questions? In asking these questions, I hope I have offered some provocation from within the Australian context for you to explore it in early childhood teacher learning in your own context and to strengthen pedagogies for equity and respect for diversity. The equity issues that arise for Australian early childhood teachers in current times that I have explored in this chapter are specific to Australia. However, I believe that early childhood teachers must learn to become critical meaning-makers, whatever their context, if they are to deconstruct the exclusionary and reconstruct the inclusionary in their daily pedagogies with young children.

NOTE

An earlier version of this chapter was presented as a keynote address titled, "Equity and Respect for Diversity: The role of early childhood professionals as critical meaning makers" to the: "*That's not fair!*" Final conference of the national dissemination project Kinderwelten, Friday November 30, 2007, Berlin, Germany.

References

Aboud, F. and Doyle, A. (1996). "Does Talk of 'Race' Foster Prejudice or Tolerance in Children?" *Canadian Journal of Behavioural Science*, 28(3), 161–170.

Ang, I. (2003). "I'm A Feminist but...'Other' Women and Postnational Feminism." In R. Lewis and S. Mills (eds.), *Feminist Postcolonial Theory: A Reader*. Edinburgh: Edinburgh University Press. 190–206.

Ashcroft, B., Griffiths, G., and Tiffin, H. (1998). *Key Concepts in Post-Colonial Studies*. London: Routledge.

Atkinson, J. (2002). *Trauma Trails Recreating Song Lines*. North Melbourne: Spinifex Press.

Brown, B. (2001). *Persona Dolls and Young Children*. London: Trentham Books.

Connolly, P. (1998). *Racism, Gender Identities and Young Children*. London: Routledge.

Cultural Diversity Resource (2006). Centre for Equity and Innovation in Early Childhood, the University of Melbourne, Melbourne.

Davis, K. (2004) "Approaches to Teaching Young Children about Indigenous Australians," Unpublished Doctoral Thesis, University of Melbourne.

Elder, C., Ellis, C., and Pratt, A. (2004). "Whiteness in Constructions of Australian Nationhood: Indigenes, Immigrants and Governmentality." In A. Moreton-Robinson (ed.), *Whitening Race*. Canberra: Aboriginal Studies Press. 208–221.

Foley, G. (1998). *The Power of Whiteness*. Retrieved February 17, 2008, from http://www.kooriweb.org/foley/essays/essay_5.html.

Friere. P. (1970). *Cultural Action for Freedom*. (Trans. Myra Bergman Ramos). Middlesex: Penguin Books.

Ghandi, L. (1998). *Postcolonial Theory: A Critical Introduction*. Sydney: Allen and Unwin.

Hage, G. (1998). *White Nation: Fantasies of White Supremacy in a Multicultural Society*. Amandale, NSW: Pluto Press.

Hirschfeld, L. (1995). "Do Children Have a Theory of 'Race?'" *Cognition*, 54, 209–252.

Human Rights and Equal Opportunities Commission (HEROC) (1997). Bringing them Home. Retrieved August 16, 2007, from http://www.austlii.edu.au/au/special/rsjproject/rsjlibrary/hreoc/stolen/.

Johnson, D. (1992). "'Racial' Preference and Biculturality in Bi'racial' Preschoolers." *Merrill-Palmer Quarterly*, 38(2), 233–244.

MacNaughton, G. (1993). "Gender, Power and Racism: A Case Study of Domestic Play in Early Childhood." *Multicultural Teaching*, 11(3), 12–15.

———. (2001a). "'Blushes and Birthday Parties': Telling Silences in Young Children's Constructions of 'Race.'" *Journal for Australian Research in Early Childhood Education*, 8(1), 41–51.

———. (2001b). "Silences and Subtexts of Immigrant and Non-immigrant Children." *Childhood Education*, 78(1), 30–36.

———. (2001c). "Back to the Future—Young Children Constructing and Re-constructing 'White' Australia." Keynote presented at the New Zealand Council of Educational Research, *Early Childhood Education for a Democratic Society National Conference*, Wellington, NZ, October 26. Published conference proceedings.

———. (2001d). "Dolls for Equity: Foregrounding Children's Voices in Learning Respect and Unlearning Unfairness." *New Zealand Council for Educational Research Early Childhood*. Folio, 5, 27–30.

———. (2005). *Doing Foucault in Early Childhood Studies.* London: Routledge.

Mac Naughton, G. and Davis, K. (forthcoming). *Race and Early Childhood Education: An International Approach to Identity, Politics, and Pedagogy.* New York: Palgrave McMillan.

Mac Naughton, G., Davis, K., and Smith, K. (2008). "Working and Reworking Children's Performance of 'Whiteness' in Early Childhood Education." In M. O'Louglin and R. Johnson (eds.), *Working the Space in between: Pedagogical Possibilities in Rethinking Children's Subjectivity.* New York: SUNY Press.

Mac Naughton, G. and Hughes, P. (2007). "Teaching Respect for Cultural Diversity in Australian Early Childhood Programs: A Challenge for Professional Learning." *Journal of Early Childhood Research*, 5, 189–204.

Martiniello, J. 2007, "Open Letter to Major Australian Newspapers." Retrieved June 19, 2007, from http://www.publicenemy.com/pb/viewtopic.php?p=205191&sid=46457d5ea377b81f5515f3432e13b1f9.

McLaren, P. (1995). *Critical Pedagogy and Predatory Culture: Oppositional Politics in a Postmodern Era*. London: Routledge.

Pownall, E. (1960). *The World We Share*. Sydney: Shakespeare Head Press Ltd.

Ramsey, P. (1991). "The Salience of 'Race' in Young Children Growing up in An All-White Community." *Journal of Educational Psychology*, 83(1), 28–34.

Ravenscroft, A. (2003). A Picture in Black-and-White: Modernism, Postmodernism and the Scene of "Race." *Australian Feminist Studies,* 18(42), November 2003, 233–244.

Taylor, P. (1993). *The Texts of Paulo Freire, Buckingham*. Philadelphia: Open University Press.

CHAPTER 12

How the Politics of Domestication Sabotage Teachers' Professional Growth and Students' Learning

Alberto J. Rodriguez

WHAT DOES THE LITERATURE SUGGEST WE MUST DO?

According to Borko (2004), "each year, schools, districts, and the federal government spend millions, if not billions of dollars on in-service seminars and other forms of professional development that are fragmented, intellectually superficial, and do not take into account what we know about how teachers learn" (p. 3). To avoid this pitfall, many educators and researchers have proposed a variety of approaches for improvement. For example, Loucks-Horsley, Hewson, Love, and Stiles (1998) suggest seven principles for establishing what they call "effective professional experiences" designed to provide teachers with research-based opportunities for professional growth. Similarly, various commissions on teaching and teacher education have outlined a variety of factors that influence what teachers teach and how they teach, and the pervasive impact these factors ultimately have on the student achievement gap and participation in science, technology, engineering, and mathematics. These commissions have also proposed specific suggestions for improvement (Glenn Commission, 2000; Mendoza Commission, 2000; the National Commission on Teaching and America's Future, 1996).

While these studies and reports are very useful, there is little research on how best to specifically support science teachers interested in making their practice more gender inclusive, inquiry-based, and socially/culturally relevant to all students—in other words, how best to support ways to move pedagogies beyond exclusion. Though there are many studies that tend to imply that teaching all children and teaching for diversity are important aspects, this body of research usually expects teachers to tackle teaching for diversity and understanding on their own, without examples and guidance on how to meet the demands of working with students from economically impoverished and culturally and linguistically diverse backgrounds. In order to address this issue, we carried out a teacher professional development research project that sought to implement what many scholars have suggested for establishing effective programs, in addition to what we have also learned from our own previous studies.

THE I²TECHSCIE TEACHER PROFESSIONAL DEVELOPMENT RESEARCH PROJECT

The main goal of the I²TechSciE Project was to establish a collaborative community of practice that was *responsive* to the teachers' professional needs, *on-site* (involving multiple visits to their classrooms), and *on going* (three-year project). During the first two years of the project, we recruited all of the grade 4 through 6 teachers (n = 9) at an economically disadvantaged and culturally diverse urban school in the Pacific Southwest. By the third year of the project, we expanded the study to two other culturally diverse schools in the same district.

To enrich the educational opportunities afforded to students in these disadvantaged schools, our project also provided teachers and their students with training and access to high-end scientific equipment and software. That is, we provided a wireless laptop cart (with ten laptops), Vernier data gathering probes and computer interphases, digital video cameras, a digital microscope, and a variety of web-based and laptop-based instructional software. We were aware that the literature on the use of learning technologies in the classroom describes many obstacles that prevent teachers from actually implementing what they learned in off-site workshops (Cuban, 2001; Pflaum, 2004). We also recognized that just providing access to sophisticated computer equipment and software to teachers and students from economically disadvantaged schools is not enough to effect long-lasting change (Cuban, 2001). Therefore, we took steps to address the commonly reported pitfalls. In keeping with the responsive design of the project, we offered teachers two-week professional development summer institutes in which we modeled how to integrate the available technologies with inquiry-based and culturally/socially relevant science activities. This pedagogical and technical support was continued during the academic year as mentioned above.

Another important aspect of our study is that we used concept maps as an alternate form of assessment to quantitatively measure students' knowledge growth. Concept maps provide a graphic representation of students' thinking and can also be used to measure the students' propositional knowledge and cognitive structures in a subject domain (Novak and Gowin, 1984). There are many variations in the use of concept maps as pedagogical and/or assessment tools, so we decided to use semi-structured, fill-in-the-blank concept maps for this project. We did not feel that any available test could truly measure the increased engagement, participation, and achievement we witnessed in the classroom or gathered during student interviews, but conducting pre-instruction and post-instruction unit tests using the concept maps enabled us to gather more information on students' learning. It is worth mentioning that these unit tests were challenging, and they required students to demonstrate hierarchical understanding. That is, they were expected to not just write a complete definition, but to also be able to properly link an example (in their own words) with that definition and/or link other related terms or components to a central concept. More details on the design of the study and on students' performance on the concept map tests, as well as the various strategies we implemented to manage the challenges we encountered, can be found in Rodriguez and Zozakiewicz (in press); Rodriguez and Zozakiewicz (2008); and Rodriguez, Zozakiewicz, and Yerrick (2005). Readers are also invited to visit our project's Web site at http://edweb.sdsu.edu/i2techscie. Multiple photo galleries, videos, and classroom resources are available there for those interested in obtaining a closer look at this project in action.

The collaborative and transformative focus of the I²TechSciE Project was guided by sociotransformative constructivism (sTc). This is a theoretical framework that merges the tenets of multicultural education (as a theory of social justice) with social constructivism (as a theory of learning) (Rodriguez, 1998). However, due to space constraints, I cannot describe here how sTc informed all the aspects of our project. Similarly, I cannot describe the quantitative and qualitative mythologies that we used to gather and analyze the multiple data sources we gathered for this study. As mentioned above, these aspects of the project are explained elsewhere. What I would like to do here instead is to take the bulk of the space available to provide an in-depth case study analysis (Patton, 1987) of one of the teachers in the project who showed the most professional growth. The case study is based on five ethnographic interviews conducted over the course of three years, weekly classrooms visits, many informal conversations, and field notes. My goal is to draw attention to Pedro's story—a grade six Latino teacher whose efforts to grow as an effective educator were truncated by top-down policies of the school district. These policies—in turn—were being implemented due to the punitive nature of the No Child Left Behind Education Act. To put it simply, this case study demonstrates how a school district itself ended up sabotaging a promising teacher's professional growth and his students' prospects of learning by engaging in a *politics of domestication*. I define this term as the negative process of acculturation by which one's ideals and commitment to grow and to work for social justice are tamed and reduced to fit dominant discursive practices (Rodriguez, 2006). I hope that this case study will provide additional insights into the causes for the pervasive gap in student achievement in science and for the slow progress we continue to make in the professional development science teachers in the United States.

From Manual Labor Working Class to Becoming a Middle-Class Teacher

Pedro is one of those unique individuals whose transparent honesty is evident in his everyday discourse. He is indeed the quintessential person who embodies the expression, "what you see is what you get." This made Pedro an excellent informant from a researcher's point of view, but he was a lot more than that. As a bilingual Latino (male), Pedro was also a great role model for his students. During multiple class visits and from multiple focus group interviews with his students over three years, it was evident that Pedro's students admired and respected him in his fatherlike role. He was like that fun uncle that many of us may have been fortunate enough to have when growing up. The kind of uncle who has a good sense of humor, honestly cares about you, and always has good stories to share, but who also could put you in your place with just one look if you get out of line.

Pedro's obvious love for teaching and for working with children developed after making a significant career change. He worked for over ten years doing intensive manual work before deciding to become a teacher. Here he recalls his first encounter with what it meant for him to transition from working class laborer to middle-class teacher:

> It was just a dinner presentation that my administrator at the time had invited me to go…It was funny, too, because it was my first year teaching…I showed up in jeans and boots, a t-shirt, and my backpack ready to take notes and stuff and everybody is in ties and, you know, I'm like, okay, formal. You had to sign in and they were gonna give you a ticket, and I remember going up to the lady, and this is my name and she kind of went, "yeah?" And I said, "I'm a teacher." [She responded,] "Oh, you're a teacher, okay, okay." They probably figured I was part of the clean up crew or something. I just remember going, "okay, just a little out of place here." (Interview I, 11)

This dinner and lecture by a scientist was the closest event to a "science workshop" Pedro had ever attended before joining our project. This was a main source of concern for him because he felt that science was his weakest area, and like most elementary school teachers in the United States he was only required to complete three general science courses to receive his teaching credentials. Furthermore, Pedro did not get a chance to practice teaching science or see other skilled teachers teach science during his professional preparation. After taking credential classes at night in the bilingual teacher education program, Pedro was quickly hired, without even completing his student teaching practicum, under an emergency hire procedure. This is not an uncommon situation given the need for bilingual teachers willing to teach in inner city schools.

Pedro's first job was at an economically disadvantaged school that he described as "all my students looked like me." Here, as the bilingual teacher, he worked for four years with mostly students whose first language was not English and who mainly came from Mexico—with only a few of them coming from various Latin American countries. Now, he was about to start his second year in the school we had chosen for our professional development project. This was the beginning of his sixth year of teaching, and he was very excited about the opportunity to actually teach science, use more learning technologies, and work with students from more culturally diverse backgrounds. He adds,

> Where as this year'll be the first year where I will actually get English only kids mixed in with bilingual kids because the program here was being revamped where they'll only have the bilingual K through 3 and if we do have new to country kids, they'll be sent to me. So,…I'll

have kids where they're not just kids from my culture now, you know. I'm gonna have a wide array of them now and, at the same time, much more polarized as far as language is concerned. (Interview I, 8)

In fact, our project's school had a student population of 56.5 percent Latino/a (Hispanic); 18.8 percent Anglo-European (white); 5.6 percent African American (black); 2.5 percent Asian; 0.6 percent First Nations (Native American); and 16 percent Other. The proportion of English Language Learners at this school was also 37 percent, and 37 percent of all students were in the free lunch program. This school had more ethnic diversity and more resources than Pedro's first school, but it was still considered to be economically disadvantaged.

Two of the features that attracted Pedro to this school was the fact that it had a computer desktop lab and every teacher had one hour a week to take students to the lab if they chose. In his own words, Pedro considered himself to be "computer illiterate" before coming to this school, and he had just begun to learn how to use them for more than word processing. He explained that since our project had an emphasis on integrating learning technologies with inquiry-based science—his two weakest areas—"…anything you guys were gonna show me, I was gonna come out ahead already and so that was really neat just playing around and learning all the different activities [during the summer institute that] you can do with computers and, hopefully, my enthusiasm towards science and math and other subjects, in general, is gonna transpose onto the kids" (Interview I, 3).

BECOMING A MORE EFFECTIVE TEACHER: INTEGRATING LEARNING TECHNOLOGIES AND INQUIRY-BASED SCIENCE IN THE CULTURALLY DIVERSE CLASSROOM

Pedro's enthusiasm and commitment to becoming a better science teacher certainly paid off. He became one of the teachers who most often (and effectively) integrated learning technologies with inquiry-based science activities in his class. This was evident during our numerous planning sessions, visits to his class, and interviews with his students. As mentioned earlier, our project's design called for modeling and demonstrating how to integrate the learning technologies available through our grant with standards-based inquiry and culturally inclusive science activities during the summer institutes. We then provided in-class support to assist teachers in developing confidence and skills as they sought to implement the pedagogical approaches being promoted by the project. This was indeed a challenging task for the participating teachers, especially for teachers such as Pedro, who had seldom used technology and inquiry-based approaches in his classroom prior to participating in this project. For example, when Pedro attempted to use a more hands-on and inquiry-based approach in his class, he relied on the Full Option Science System (FOSS) kits that the school district sometimes made available to teachers. However, he explained that most often he would have students read from the science text and answer questions. While he wanted to do more hands-on and minds-on activities—like many elementary teachers in the United States—he lacked the content knowledge and pedagogical skills to develop and implement curriculum to fit the unique sociocultural contexts of his classroom (Weiss, Pasley, Smith, Banilower, and Heck, 2003). One of our project's main goals was to address this need and provide Pedro with the content knowledge, pedagogical skills, and equipment necessary to meet his professional development goals.

This is how he described the progress he made after being involved in the project for two solid years,

> When I started the program, and when you guys first came to the staff meeting, I mean, this is my opportunity to learn both. And I knew I wanted to teach science. I just didn't know how to teach it and make it interesting and fun instead of just right out of a book, you know? So yeah, in those two areas and without a doubt, there's no question that I have learned a lot. (Interview IV, 9)

Pedro went on to explain that his confidence level and content knowledge have also increased to a level at which he felt he could enhance other lessons. He explains,

> [I'm] mostly just self-reflecting on my lessons that I've given before, starting at square one technology and science wise, and three years down the line now, I can now go back at some of those lessons and hey, I can improve this and this way. I can do this better now. In that sense, I've definitely noticed a difference. But yeah, you can build and improve on the lessons that you've probably fell flat on your face the first time around. (Interview IV, 10)

Of all of the teachers participating in the project, Pedro was the one who showed the most significant and consistent improvement, and the one who best understood the transformative design of our professional development project. In other words, it was clear to him that our purpose was to provide ongoing, responsive, and on-site support as needed and then gradually become less involved in their classes so that the participating teachers could more independently implement the integrated curriculum. He demonstrates his insight into the design and ultimate goals of the project here,

> ...With us being in this [project] for the third year now and by us being able to be more comfortable and confident in our room, I think we're finally at the core of what's supposed to be happening here is that you guys are now allowed to actually do your job and just come in, observe and collect data...It's like everybody's really starting to fall into their roles. We teach not just a concept, but the technology as well. And now you guys as professors can just come in, observe, help whenever help's needed, but now you have more time for the documentation stage [of the project]. (Interview IV, 11)

It is important to stress that although our project had a positive impact on Pedro and on the other participating teachers, we did encounter many challenges, and the observed improvement in teachers' practice was not uniform across all participants. As expected in the complex context of working with individuals with different personalities and multiple demands on their personal and professional lives, there were many instances in which the participants' espoused beliefs and commitments to improve their practice that did not equate with their action. This fact makes Pedro's story even more compelling because he was one of the teachers from our study who was indeed making continuous progress. For example, Pedro, and several of the other teacher participants, had difficulty implementing some of the culturally inclusive activities we had either modeled during the summer institutes and/or planned together during our unit planning sessions. In the following excerpt, Pedro explains how he began to see a more pronounced difference in the participation and engagement of his female students when he more purposely followed some of the gender inclusive strategies we had previously modeled and/or discussed:

> Maria, never volunteers to answer, you know, we're talking about female scientists and stuff and all of a sudden I have twice as many, three times as many girls' hands in the air than I do boys'. These girls are, oh God, you know, I'm trying to keep the smile off from my face and keep the lesson going as I'm sitting in front of the class because you're watching this unfold in front of the class while you're giving a lesson and you're just, you know, you just kind of want to turn

around and just say, "Yeah." Okay, back to what we were talking about girls. I see it [this confidence] with the girls when we play soccer too, where some of these girls start, where they finish, just the way they carry themselves around school sometimes, the body language alone, and to begin to see that also in the classroom, it's, God, it's fantastic. (Interview V, 19)

The increase in students' self confidence that Pedro noticed also encouraged us to take advantage of this opportunity to assist teachers in implementing the pedagogical strategies and learning technologies being promoted by the project. In other words, since this was a longitudinal project, participating teachers started to notice that many of the students who came from the lower grades (i.e., fourth or fifth grades) had already accumulated a great deal of knowledge by the time they got to the next grade, they also showed confidence and skills in the use of the available learning technologies. We also noticed that other participating teachers (and even nonparticipating teachers from the lower grades) often "borrowed" a pair of students from Pedro's class to help them in with the wireless laptops, Vernier probes, and other equipment or software, or to just simply help them in the school's computer lab. In addition, many students began to develop "short cuts" and more efficient ways to set up and conduct laboratory activities with the Vernier probes that we had. Therefore, we decided to expand on this phenomenon for the benefit of all teachers and develop the intervention strategy of *students as change agents* (Rodriguez and Zozakiewicz, 2008). This approach consisted of providing specialized training to a cohort of students from each of the participating classrooms (grades fifth to sixth) so that they could become *Tech Wizards* (in charge of assisting their teachers with the available learning technologies) and *Tech Coaches* (in charge of helping with their peers with all aspects of the activity). This strategy proved to be very effective (and popular) with the teachers and the students. It provided more support to teachers (especially the grade four teachers who had to start right from the beginning with all aspects of the learning technologies and pedagogical strategies being suggested), and the students indicated in interviews how much they enjoyed playing this role.

IMPACT ON STUDENTS' ACHIEVEMENT

In addition to increased confidence, knowledge, and skills in the use of sophisticated learning technologies to gather, analyze, and present scientific knowledge, Pedro's students also showed significant growth in science content knowledge. Students demonstrated this achievement in the specially designed unit tests on the concept maps we developed with each teacher. During interviews and informal conversations with students, they stated that they found these tests very challenging. Students would often make comments like, "I don't like these tests because I can't guess the answer." In other words, students were aware that if they did not know the answer in multiple choice tests, they always had a 25 percent chance of guessing the answer correctly. Other students would make comments like, "You really have to show what you know because you have to write the answers." In these tests, students were required to write full sentences, provide definitions and examples, and most importantly make hierarchical connections between related scientific concepts. The students received more points when they correctly linked one concept with another as required, thus showing a deeper understanding.

What is remarkable is that even though these tests were more challenging than multiple choice tests, many students performed very well. This is particularly remarkable when one considers the complexity of new vocabulary and difficulty of new concepts students are expected to learn in science. For example, the grade six geology concept map unit test we developed with the teachers on "How the Earth Changes" includes complicated terminology on plate tectonics (e.g., transformative, convergent, and divergent boundaries). For this test, the average class score was 48.56 percent, with 44 percent of all students obtaining a score of 52 percent or higher. We also conducted a dependent t-test analysis of the pre- versus post unit concept maps tests, and we found a statistically significant difference in mean scores of -43.8 percent ($p < 0.001$; t -14.28; df = 24; n = 25). This shows that most students significantly gained new science content knowledge by the completion of the unit; however, we feel that this test—just like all two-dimensional assessment instruments—does not show the depth of students' understanding and engagement when working on inquiry-based projects using learning technologies. We need to develop better tools to capture and assess the multiple modalities that students activate to complete these tasks successfully.

"Pulling the Rug:" How to Sabotage Teachers' Growth and Students' Learning

By the third Summer Teacher Professional Development Institute (the beginning of Year III of our project), we decided to expand the study to two other elementary schools. Since Pedro was one of the teachers who had shown the most growth in confidence, content knowledge, and skills, we asked him to lead one of the workshops for the new teachers. Even though this was the first time that Pedro had taken on such a role, the new participants commented that they found his workshop very useful. Indeed, he provided them with a very effective hands-on and minds-on workshop rife with actual examples of students' work and clear suggestions for avoiding pitfalls when using learning technologies in their classrooms.

Given Pedro's new teacher leader role, we asked him to head the planning and lesson development for a new unit with the other grade six teacher who had joined the project during the summer institute. Pedro was very excited about developing this standards-based grade six unit on energy because it was one of those content areas that he felt he needed to improve. Just like we had done in the previous two summer institutes, we provided teachers with time and professional support to develop their units and worked closely with them to integrate the available learning technologies with the science content and with culturally responsive and socially relevant pedagogical approaches.

Unfortunately, just when Pedro was about to start the new unit in the fall semester, a new directive came from the school district that derailed his plans. This is how Mr. Lopez, the school principal who was very supportive of the project from day 1, and who even participated during the entire first summer institute with the teachers, explained the new district policy,

> Because our English Language Learners were on the border line, or on the threshold of not passing in the area of English Language Arts, when I met with my boss [the District Superintendent], he said "you know, we will not have another school that is gonna be program improvement. I promised the [School] Board that we will have no more, because we have like 5 or 6 right now. Be sure that you don't go to program improvement." (Interview II, 11)

By "program improvement," the Superintendent meant the classification assigned to schools using the No Child Left Behind National Education Act when the students' standardized test scores fall below the expected minimum score. Therefore, a district-wide policy was implemented that consisted in increasing the time designated for English Language Development for *all students*. In this way, students were to be provided with more structured reading and writing activities geared to help them improve their scores on the standardized tests. However, since science and social studies are subject areas that are not tested at every grade,[1] like mathematics and language arts, according to the No Child Left Behind Act, these subjects were essentially expendable. Therefore, Pedro was required to drop his science instruction time from one hour almost every day to thirty to forty-five minutes per week. This drastic change in policy really frustrated Pedro and his students because he felt like his hands were tied.

> It's been frustrating, you know, because just as you're starting the get the technology and the concepts and everything, you know, everything is starting to click. You're starting to really improve on your lessons from the year before…and then, it's like the rug is pulled out from under you. (Interview V, 3)

We were frustrated too because the students complained to us that they were not doing much science or projects anymore, and all the work that Pedro put in collaboratively developing the new unit on energy essentially was going to waste. In fact, Pedro's students scored the lowest on this unit than in any other of his previous units. The average score on the unit concept map test was 24.82 percent. When we conducted the dependent t-test analysis between the pre- and post unit concept map tests, we found the difference to be -23.09 percentage points ($p < 0.001$; t = -7.56; df = 22, n = 23). While there was some knowledge growth, only two out of the twenty-three students in the class received a grade of 50 percent or above. This was demoralizing for Pedro and his students because they felt that they could have done better if they actually had the chance to complete the planned activities and projects associated with the unit.

Ironically, the devastating effects of this new policy on Pedro's teaching practice and on his students' science learning was occurring when he, in collaboration with other project teachers, had managed to successfully apply for their own school's wireless computer cart (with thirty laptops. Ours had only ten!). Helping teachers build capacity and secure their own equipment was one of the goals of our project, because we knew that once the study was completed all the equipment was to be returned to the university. However, now that the participating teachers had more laptops, they were expected to do less science.

Pedro's story could be that of any of the teachers' who are really committed to improving his/her science content knowledge and pedagogical skills. Yet, his efforts were truncated by the imposition of a policy that seems more focused on blindly improving scores on standardized tests at the expense of other curriculum subjects than in supporting teachers' professional growth and students' learning for understanding. What are the implications then for the continuing calls for teacher education reform and for the need to enhance students motivation and achievement in science?

The Politics of Domestication: Implications for Teacher Professional Development, Research and Practice

One cannot help but wonder how many more committed teachers like Pedro face the kind of *opp(regre)ssive* policies described here. That is, these polices are oppressive and regressive simultaneously. They are oppressive because they have imposed a directive on teachers (and principals) without regard to their professional knowledge and without regard to whatever programs may have already been in place to help students learn. They are oppressive because instead of utilizing the teachers' and principal's professional and craft knowledge to generate an intervention that may work best in their own contexts, the school district mandated a sweeping policy with a clear punitive consequence. As Principal Lopez exclaimed, "Basically, if things go into a negative situation [meaning if the ELL's scores did not improve], then I'm looking for a job somewhere, basically" (Interview II, 12). The school district's policies were also regressive because they contradicted what the research literature indicates about effecting positive and long-lasting change in schools (Loucks-Horsley, Hewson, Love, and Stiles, 1998; Johnston, Brosnan, Cramer, and Dove, 2000). It is important to note that this policy affected *all* students—not just the English Language Learners. In other words, even the native English speakers and the fluent, bilingual students were required to participate in the designated English Language Development periods at the expense of being provided with the regular amount of science (and social studies) instruction as required by the State and National Science Education Standards.

In short, the school district's opp(regre)ssive policy is a good example of the politics of domestication that plague education in the United States and continue to obstruct any serious effort to answer the calls for reform. With this new policy, the school district sought to domesticate and subjugate every teacher and principal—regardless of their unique working contexts, their progress in teacher professional development, and/or the differences in students' achievement or language skills, The school district's sweeping policy's sole purpose was to comply with another policy that is mainly guided by punitive accountability—the National No Child Left Behind Act.

While everyone would agree that we have already too many schools designated as "program improvement," and too many that have already undergone reorganization due to continuing low performance, we must seriously reflect on the long-term socioeconomic, cultural, and moral impact of just simply cutting instructional time in science to give more time to language instruction as if these subjects were just like deck chairs on the sinking Titanic. This rearranging may provide a false and temporary sense of repose, but it is not going to prevent the ultimate and unavoidable fate that is bound to follow educational policies that focus on quick fixes when it is the whole ship that needs repair. If we are serious about implementing long-lasting and effective changes in our schools, and if we are to reap the benefits from the millions of dollars invested in educational research to enhance students' engagement and achievement in science, we then need to pay more close attention to whatever ongoing and successful efforts may be taking place at each and every school before sweeping and domesticating policies truncate these efforts. After all, unlike Borko's (2004) lament cited at the beginning of this chapter, the I²TechSciE Project's responsive, on-site, and longitudinal design was having a positive and transformative impact on most of the teachers and their students. Unfortunately, it was the school district's opp(regre)ssive polices that negated the efforts of a dedicated teacher like Pedro who—after three years of continuous professional growth—found himself again frustrated with and demoralized by his administrators, "Yeah, you can only bang your head against the wall for so long before you start to feel the bruise" (Interview V, 8).

NOTE

1. Science standardized tests are administered in grade 5 only.

REFERENCES

Borko, H. (2004). "Professional Development and Teacher Learning: Mapping the Terrain." *Educational* Researcher, 33, 3–15.

Cuban, L. (2001). *Oversold and Underused: Computers in the Classroom.* Cambridge, MA: Harvard University Press.

Glenn, J. (2000). *Before It's too Late: A Report to the Nation from the National Commission on Mathematics and Science Teaching.* Washington, DC: US Government.

Johnston, M., Brosnan, P., Cramer, D., and Dove, T. (eds.) (2000). *Collaborative Reform and Other Improbable Dreams: The Challenges of Professional Development Schools*. New York: State University of New York Press.

Loucks-Horsley, S., Hewson, P., Love, N., and Stiles, K. (1998). *Designing Professional Development for Teachers of Science and Mathematics*. Thousand Oaks, CA: Corwin Press Incorporated.

The Mendoza Commission (2000). "Land of Plenty: Diversity as America's Competitive Edge in Science, Engineering and Technology." Report of the Congressional Commission on the Advancement of Women and Minorities in Science, Engineering and Technology Development. Washington, DC: Available with Author.

The National Commission on Teaching and America's Future (1996). *What Matters Most: Teaching for America's Future*. New York: Available with Author. http://www.tc.colimbia.edu/~teachcomm. Retrieved August 8, 2007.

Novak, J. D. and Gowin, D. B. (1984). *Learning How to Learn*. New York: Cambridge University Press.

Patton, M. Q. (1987). *How to Use Qualitative Methods in Evaluation*. Newsbury Park, CA: Sage.

Pflaum, William D. (2004). *The Technology Fix: The Promise and Reality of Computers in Our Schools*. Alexandria, VA: Association for Supervision and Curriculum Development.

Rodriguez, A. J. (1998). "Strategies for Counterresistance: Toward Socio-transformative Constructivism and Learning to Teach Science for Diversity and for Understanding." *Journal for Research in Science Teaching*, 35(6), 589–622.

———. (2006). "The Politics of Domestication and Curriculum as Pasture in the United States." *Teaching and Teacher Education*, 22, 804–811.

Rodriguez, A. J., and Zozakiewicz, C. (in press). "Facilitating the Integration of Multiple Literacies through Science Education and Learning Technologies." In A. J. Rodriguez (ed.), *Science Education as a Pathway to Teaching Language Literacy*. Rotterdam, Netherlands: SENSE Publishing.

Rodriguez, A. J., Zozakiewicz, C., and Yerrick, R. (2008). "Students Acting as Change Agents in Culturally Diverse Schools." In A. J. Rodriguez (ed.), *The Multiple Faces of Agency: Innovative Strategies for Effecting Change in Urban School Contexts*. Rotterdam, Netherlands: SENSE Publishing.

Weiss, I. R., Pasley, J. D., Smith, P. S., Banilower, E. R., and Heck, D. J. (2003). *A Study of K-12 Mathematics and Science Education in the United States: Looking inside the Classroom*. Chapel Hill, NC: Horizon Research, Inc.

CHAPTER 13

Rethinking American Education through Indigenous Wisdom and Teachings

Kathryn D. Manuelito and Maxine Roanhorse-Dineyazhe

The most effective school pedagogies have been revealed by the study of Native American classrooms that use their traditional cultural patterns of activity and interaction.... The communities of Native Americans, uniquely, can be the seedbed for infusing fundamental human processes of teaching/learning into public education.

— *Tharp, R. G., 2006.*

It is the responsibility of the teacher or educator to show how dominant forms of knowledge and ways of knowing constrict human capacities.

— *Ng, R., 2003.*

After 400 years of imposed and mandated Americanized schooling, indigenous youth in the United States have suffered consequences that are evident in their social relationships and academic performance. The competing ideologies and messages of cultural pluralism and assimilation have negatively impacted the daily schooling experiences of approximately 1.2 million indigenous youth (Swisher and Tippeconnic, 1999) across the United States who mostly attend public schools (Indian Nations At Risk Report, 1990). Since the Meriam Report of 1928 to the present, standardized testing has unfairly labeled indigenous youth. Reyhner and Eder (2004) have asserted

"...the greatest danger facing Indian education...is the push for outcomes assessment, state and national standards, and the associated increased use of high stakes testing in all facets of education..." (p. 11). Conflicting expectations of assimilation and diversity as well as biased assessments have marginalized indigenous youth and have contributed to a self-fulfilling prophesy for both indigenous students and the teachers who teach them.

American Indian Nations have been, and continue to be, concerned about improving learning outcomes of their youth. Instead of blaming youth and the experiences they bring to school as in the deficit theory, which is the philosophy behind most programs designed to improve academic outcomes for indigenous youth, educators are reexamining ways indigenous youth experience schooling, specifically the pedagogy they receive. Some Native teacher education programs are critically reconsidering the methods and underlying philosophies of instruction in order to improve schooling experiences for indigenous children. One such program is the Indigenous Teacher Preparation (ITP) Program at Arizona State University, Tempe, Arizona. This program has developed curriculum that infuses teachings, values, and histories of indigenous peoples. It emphasizes the importance of relationships and place for capable, effective teaching. This chapter will describe the ITP program, which goes beyond mainstream pedagogies and is considered to be one whose purpose is to transform indigenous communities. Before discussing ITP at Arizona State University, three major topics will be considered: first, the context of indigenous youths' lives; second, indigenous epistemology that is infused into ITP curricula; and third, the rationale for developing more indigenous teachers.

Context of Indigenous People's Lives

Since the 1819 Civilization Act, which funded missionaries to proselytize and educate in the Euro-Western tradition, American educators have generally ignored the contexts of indigenous students' lives. Since that time, indigenous people have been coming to terms with the United States government's full-scale disruption of their lives. Thornton (1998), an indigenous scholar, refers to these experiences as a "holocaust" that is "crucial to understanding the full impact of colonialism upon Native Americans and the social, cultural, biological, and perhaps psychological changes they subsequently underwent" (19). Not only was the population of indigenous people reduced since initial contact 500 years ago, through various immoral schemes, from approximately 9.8 million to perhaps 375,000 around 1900 (Thornton, 1998, 19), but indigenous children as young as three years old (Child, 1995, 110–115) were forced to attend boarding schools far from their homes and communities. Today, being the smallest minority group in the United States, the American Indian/Alaskan Native as indigenous people are approximately 2.5 million or 4.1 million in combination with another race (U.S. Census Bureau, 2001a, 2001b). A major language shift is also occurring among the 552 federally recognized tribes. Among these tribes whose languages and cultures are different from one another, 154 languages are still spoken, 45 are on the verge of extinction, and another 90 are projected to disappear by the year 2050 (Crawford, 1995; Karauss, 1996; cited in Francis and Reyhner, 2002, 3). Federal and State mandates have threatened indigenous identities that are expressed through language and culture. The emphasis on English Only and past Federal Policies have created a cultural and linguistic crisis for indigenous peoples in the United States.

The generational trauma of institutionalized child development has impacted indigenous communities in innumerable ways. Duran and Duran (1995) refer to trauma inflicted upon indigenous people in the United States as the "soul wound" (24) that indigenous people are presently resolving, one of which is reexamining the complexities of Euro-American education. In the lives of indigenous people, education remains a paradox. While some indigenous tribal Nations, especially college-age youth, voice strong desires to maintain their sovereignty, cultures, and languages through formal education (Tierney, 1991; Brayboy, 2004), others are seeking alternatives to this education by infusing their epistemology into the methods of mainstream pedagogy.

> Schooling for indigenous students remains a daily contradiction when they are taught from their home and communities to be proud of their language and culture but are expected to leave this pride at the door as they enter their classrooms where only the White American culture and the English language are honored. Schools have become places where indigenous youth learn to conform to mainstream ways, rather than places of opportunity and access. Multicultural education and the honoring of diversity seem to be operational only during "Thanksgiving" and "Columbus Day," events which have further alienated indigenous youth by reinforcing myths propagated by slanted historical reporting. American educators have contributed both intentionally and unintentionally to the unsuccessful schooling experiences of indigenous youth. (Loewen, 1995, Chapters 2 and 3)

Searching for solutions to these perplexing and frustrating circumstances, indigenous elders, professionals, and parents have focused attention on the predispositions of the teachers of their children. "Teacher education has never been more important for Indian Country" (Ambler, 1999). Boyer (2005) states that to improve teacher education for indigenous youth, "Native educators need to focus their resources on transforming tribal communities." Communal goals are foremost in indigenous communities, unlike the value of individualism in the Euro-Western society. That is why many indigenous youth will state "to help my people" as the reason for attending higher education. Teacher education for indigenous preservice teachers must include communal goals of the unique community in which preservice teachers will be teaching. The next section will discuss the indigenous epistemology that is central to a transformative teacher education program that has the purpose of transforming tribal communities. This epistemology is infused into the ITP curricula.

Indigenous Epistemology

Battiste and Henderson (2000), indigenous scholars from Canada, have identified six characteristics of indigenous epistemology. They have summarized these characteristics in the following statement: "Perhaps the closest one can get to describing unity in Indigenous knowledge is that knowledge is the expression of the vibrant relationships between the people, their ecosystems, and the other living beings and spirits that share their lands" (p. 42).

Similarly, Cajete (1994), a Tewa scholar from New Mexico, describes indigenous epistemology:

> Environmental relationship, myth, visionary traditions, traditional arts, tribal community, and Nature centered spirituality have traditionally formed the foundations in American Indian life for discovering one's true face (character, potential, identity), one's heart (soul, creative self, true passion), and one's foundation (true work, vocation), all of which lead to the expression of a complete life. (p. 23)

Another esteemed scholar and philosopher is the late Vine Deloria, Jr., who defined indigenous epistemology more succinctly by stating that the scientific concepts of space, time, and energy are irrelevant in the Indian world. Instead, the concepts of power and place and the relationships between the two that create personality are experiential dimensions that have meaning in the Indian world:

> ...place and power, the latter perhaps better defined as spiritual power or life force. Familiarity with the personality of objects and entities of the natural world enabled Indians to discern immediately where each living being had its proper place and what kinds of experiences that place allowed, encouraged, and suggested. And knowing places enabled people to relate to the living entities inhabiting it. (Deloria and Wildcat, 2001, 2–3)

Other researchers describe indigenous epistemology through worldview, "Most Indigenous people's worldviews seek harmony and integration with all life, including the spiritual, natural, and human domains" (Burger, 1990; Knudtson and Suzuki, 1992; cited in Kawagley, 1995, 2). Indigenous epistemology defines identities, history, and spirituality of the indigenous people.

Indigenous worldview and indigenous epistemology are incompatible with the Western worldview. This incompatibility has extended into the educational system or schooling. "The Western educational system has made an attempt to instill a mechanistic and linear world view in indigenous cultural contexts previously guided by a typically cyclic world view" (Kawagley, A. O., 2001, 1). Deloria accounts for the problematic and perplexing experiences of indigenous students:

> Education in the English-American context resembles indoctrination more than it does other forms of teaching because it insists on implanting a particular body of knowledge and a specific view of the world, which often does not correspond to the life experiences that people have or might be expected to encounter. (Deloria and Wildcat, 2001, 42)

The incompatibilities of epistemologies and worldview frame the often frustrating and perplexing schooling experiences for indigenous youth.

RATIONALE FOR
NATIVE TEACHING AND TEACHERS

The 1990 Indian Nations At Risk Report (INAR) (Cited in Reyhner and Eder, 2004, 10) indicated that nonindigenous teachers often do not want to teach indigenous students, thus creating an unfriendly school climate for indigenous youth. Zumwalt and Craig (2005) have stated that in the United States "most teachers are female, White, and monolingual" (138) and furthermore "...the supply of minority teachers is so small that even students in predominantly minority schools usually have mostly White teachers" (145). Problems arise when students and their teachers come from different backgrounds. "White mainstream teachers tend not to have the same cultural frames of reference and point of view as their diverse students and may have difficulty serving as role models or cultural brokers between home and school" (Goodwin, 2000; Villegas and Lucas, 2002; cited in Hollins and Guzman, 2005, p. 482). Most importantly, indigenous students do not have equal access to education when their teachers do not make the effort to learn about them and, as mentioned above from INAR, don't even want to teach them. "Unless prospective teachers have opportunities to rethink and change their attitudes and beliefs, the students who are in the greatest academic need may also be the ones least likely to have access to rich learning opportunities" (Gay, 1993; Villegas and Lucas, 2002; cited in Hollins and Guzman, 2005, p. 482). Presently, few teacher education programs provide appropriate information and experience for preservice teachers of indigenous youth. The relationship between student and teacher is of utmost importance for the learning outcomes of the student.

More than twenty years ago, Tippeconnic (1983) reported that the most important relationship in the educational system occurs between teacher and student, and this relationship directly impacts the learning process for Native youth. Cummins (2003) views the interactions between educators and students (termed micro-interactions) as the most immediate determinant of student success or failure in school. He states: "Only teacher-students interactions that generate maximum identity investment on the part of students, together with maximum cognitive engagement, are likely to be effective in promoting achievement" (p. 51).

Research has demonstrated that teachers of color contribute to their student populations by creating opportunities to engage in culturally congruent pedagogy, serving as role models, reinforcing the students' cultural identities, and fostering collaborative efforts with parents and communities (Foster, 1994; Darling-Hammond, 1999; Gay, Dingus, and Jackson, 2003; Ladson-Billings, 2000). "Teachers of color who share the same cultural background as their students make a special contribution through their insider knowledge of culture and language" (George, 2005, p. 2). Delpit (2005) suggests that "appropriate education for poor children and children of color can only be devised in consultation with adults who share their culture" (p. 16). Issues related to cultural differences and learning styles of American

Indian students, as well as the influences of Native languages and cultural programs on academic performance, require Native teachers.

To provide successful learning experience for indigenous students, teachers will have to actively construct new information upon the experiences of their students, which include culturally influenced learning behaviors, communication styles, and values, in addition to the histories and epistemology of Native learners. Teachers who work with Native children must possess knowledge and have respect for the pupils they teach. Teacher education programs must critically take an in-depth look at the programmatic aspects such as curricular issues as well as recruitment policies and procedures of candidates preparing to be teachers (Demmert, 2001; Gajar, 1985; NCTAF, 2004; Swisher, 1997).

Most postsecondary mainstream institutions (i.e., colleges and universities) provide teacher education programs focusing on mainstream educational practices. Educational experiences in these programs prepare teachers to teach within the mainstream contexts. The majority of these postsecondary institutions do not provide teacher education in the area of Indian education. Teacher education that focuses on developing a teaching force for indigenous communities began only recently. "In the last 30 years, tribal universities, and state universities and colleges have provided Native teacher preparation programs" (Manuelito, 2003, p. 3). Reyhner (1992) cautions against assumptions that all Native teacher preparation programs are centered on Native epistemology and decolonizing methodologies. For instance, many programs maintain the curricular aspects of mainstream teacher education programs. The only thing Native about these programs is perhaps the word "Native" in the name of the programs. With this in mind, the Indigenous Teacher Preparation Program at Arizona State University was developed and established to provide an alternative to mainstream teacher education and to reconceptualize indigenous teacher education.

Indigenous Teacher Preparation Program

The Indigenous Teacher Preparation Program (ITP) was established in 2004. ITP has a multi-tribal outreach and has had enrollment from Pima, Tohono 'Oodham, Gila River, Cocopah, Seminole, Yaqui, Diné (Navajo), Apache, and Hopi tribes, tribes with diverse languages and cultures. Traditional or regular teacher education programs, though addressing diversity, do not specifically address Indian education and all the intricacies and impacts of tribal histories, social interactions, and cultural traditions. These institutions offer teacher training in Indian education only in a limited way and only when federal funding is granted. Therefore, as grants are depleted or are no longer available, the training programs cease to exist. However, "the traditional problems in Indian education which have been perpetuated since colonial times are beginning to find meaningful solutions through the bold innovations fostered by [colleges and universities]" (Clark, 1972, p. 3). The ITP program is an example of a bold innovation because it has been permanently incorporated into Arizona State University's College of Education teacher education program and is not dependent upon federal funding for its existence as a separate federally funded program.

One essential priority in Indian education is to have caring and effective teachers. Teachers must be knowledgeable, sensitive, and responsive to the histories, cultures, and language issues of Native communities. They must value the whole child without looking upon them with the deficit lens that many Native children are looked upon everyday. This important relationship between teacher and students is modeled in the ITP by indigenous instructors for Native preservice teachers. This is a significant difference in ITP as compared to many Native teacher education programs nationwide. The core teaching force of ITP is mainly indigenous professors who are full-time faculty of Arizona State University.

When ITP was established, the vision of a unique teacher education program was shared among many individuals, faculty, students, and community members. The initial planning process began with a prayer, and the seed for ITP was planted. The process for program development was cultivated by first creating a mission statement. The mission guided every aspect of the program, from the program model to its curriculum to the participants and finally to the resources and support systems.

ITP was established as a cohort model. This approach has been found to be effective. "Researchers found that candidates reported that the cohort group increased opportunities for students to learn from each other, created a comfortable academic climate, and sustained a sense of group belonging" (Hollins and Guzman, 2005, 505). In addition, researchers found that "when placed in a separate cohort, candidates of color developed a sense of community, a stronger sense of ethnic identity, and an affirmed commitment to working for social justice" (Hollins and Guzman, 2005, 505). For ITP the sense of belonging to a community was extremely important. One student commented at an event, "I'm in ITP. ITP is like my second family. I like that feeling." Another aspect in the program model is the flexible scheduling. Many Native teacher education candidates are "non-traditional" students, which means that they often have additional responsibilities such as caring for their children or caring for family members. A mixture of day and evening classes was most preferred and scheduled in that manner.

The ITP curriculum incorporates an overarching emphasis on indigenous education. For example, the course "History of American Indian Education" examines the relationships among the political economy, the prevailing ideology, and the educational practices that influenced federal policies and schooling for Native youth. This course helps students evaluate the competing ideologies of cultural pluralism and assimilation, the representation of Native people in school curricula, and issues of self-determination, identity, and language maintenance. In another course, "Issues in Language and Literacy of Indigenous People," students are encouraged to think critically about the serious language shift occurring in indigenous communities. They explore methodologies and strategies that encourage revitalization and maintenance of language and culture in indigenous communities, which is basic to the revitalizing and strengthening of identity. The course "Foundations of Instruction for Indigenous Classrooms" is critical for preservice teachers who will be teaching Native children. Decolonizing methodologies that emphasize the utilization of indigenous resources and teachings are stressed. Other methods courses have been taught focusing on indigenous communities. Furthermore, field experiences are provided in schools with a significant population of indigenous students. For example, ITP has partnered with nearby Native communities for access to schools in order to apply and practice teaching experiences gained in methods courses. Therefore, to ensure the vital nature of the program, all methodology courses are prefaced with Indian Education notations. Notably, and most importantly, ITP had mostly Native instructors as mentioned above. Where courses could not have Native instructors, it was necessary for instructors to have extensive experience with Native learners and communities. Program staffing continues to be made with caution. For example, Native instructors who perceive Native children from a deficit perspective only perpetuate the inappropriate schooling experiences that have led to underachievement.

In ITP, recruitment and retention of program participants is a significant and serious responsibility. The phrase "anyone can teach" creates a "slippery slope" that demeans the quality of teacher preparation programs and the candidates they produce (NCTAF, 2004, 4).

Unfortunately, all too often, teacher education becomes a last resort for students who couldn't make it in other majors. For example, a student who finds that a business major is too difficult changes to an education major and maintains the idea that drill is the most effective way to teach mathematics. Granted, mathematical knowledge is important for preservice teachers in whatever way they acquire it, but the unwillingness to try out other strategies impedes the motivation and enthusiasm that is especially needed in Native classrooms where drill is already the common nonproductive practice. Mathematical concepts exist in indigenous communities in every aspect of life. Native students have the opportunity to gain understanding and thus learn by utilizing processes from familiar settings of their homes and communities if their teachers can employ these community-based methods.

Identifying resources is a challenge in ITP because the majority of teacher candidates are not like the traditional, young college student. Most preservice teachers contend with child-care issues or issues of transportation to and from the reservations where family obligations are the norm. Support networks are critical in helping preservice candidates achieve their academic goals and fulfill their dreams of becoming teachers. ITP has implemented programmatic strategies for continuous improvement.

The most important impact of ITP was illustrated in a recent event hosted by the Gila River Tribe for ITP students on February 8, 2007. ITP students described their experiences and notably emphasized that the program provided for them an understanding of their identities. Their identities had been ignored throughout their schooling, so the information about them as indigenous people was new and empowering. Several of them mentioned that this factor made the difference for a healthy holistic approach to their schooling, which they want to impart to their future students. One student mentioned that you can have academics, but without the empowerment of identity, indigenous students cannot have access to the understanding of their unique and beautiful heritage as the foundation and bridge to successful experiences. This example reinforces Cummins (2203) statement "Only teacher-students interactions that generate maximum identity investment on the part of students, together with maximum cognitive engagement, are likely to be effective in promoting achievement" (51).

Like all living things in nature, change is inevitable and often positive. As the growth of ITP continues, its spirit is nurtured in the successes of its offspring, consisting of its graduates. Teachers who are prepared with a specialized knowledge base to meet the unique academic and cultural needs of Native learners can improve the learning outcomes of our most sacred possessions, our children. Native teachers act as the catalyst in the transformation of their communities.

Conclusion

"As Fanon and later writers such as Nandy have claimed, imperialism and colonialism brought complete disorder to colonized peoples, disconnecting them from their histories, their landscapes, their languages, their social relations and their own ways of thinking, feeling and interacting with the world" (Smith, L. T., 1999, p. 28). One approach for restoring order to the lives of Native Americans in the United States is to dismantle the contradictions inherent in formal education today. This chapter has presented a teacher education model that combines indigenous epistemology within the Euro-Western tradition of teacher education curricula to work out contradictions existing in most indigenous youth's experiences. Teacher education programs similar to and based on the ITP model will positively impact the contexts of indigenous students in their communities and at school.

The Indigenous Teacher Preparation Program (ITP) is a transformative program that recognizes that improvement for indigenous youth begins and ends with attention to and inclusion of their daily contexts. Major systemic issues, however, remain in the development and delivery of a program such as ITP, which have become all too apparent. A particular area of concern is the assimilationist curriculum that is guided by state standards and high stakes testing required by the No Child Left Behind legislation, which ITP graduates will have to grapple with in the schools that they will be teaching in. Public school curricula, even in public schools on Indian reservations, have become restrictive, and most teachers teach for the tests to ensure security to their careers. Other major systemic issues encountered by indigenous teacher educators are: How will Native people, from quasi-sovereign Nations, maintain their identities as teachers and students in American public education? How will the unfriendly climate of schooling change with so few change agents (Native teachers)? These poignant questions underscore the reality of indigenous teachers who have graduated already and are struggling within a colossal public school system.

Indigenous teacher education programs may be small and few in number compared to the monolithic teacher education programs of mainstream society, but they present an alternative to the status quo that has not worked for their youth throughout the past 400 years. ITP has also shown that "It is the responsibility of the teacher or educator to show how dominant forms of knowledge and ways of knowing constrict human capacities" (Ng, R., 2003). With the increasing events of exclusion in the social and political economy of the United States, educators must take an analytic gaze toward the indigenous peoples of this land who continue to respect, in a spiritual manner, all *relationships* to each other and all animate and inanimate beings of this universe. This chapter challenges all teachers and teacher educators to think closely about how they might serve as allies to transform the present context for indigenous children through the ITP model as well as enhance education that honors *relationships* for all youth, indigenous and nonindigenous.

References

Ambler, M. (1999). "Instilling Dreams, the Promise of Teacher Education." *Tribal College Journal of American Indian Higher Education*, 11(2), 6–7, (winter).

Battiste, M. and Henderson, J. Y. (2000). *Protecting Indigenous Knowledge and Heritage*. Saskatoon, Saskatchewan, Canada: Purich Publishing Ltd.

Boyer, P. (2005). "To Be, or Not to Be? TCUs Probe Identity Questions as They 'Indigenize' Their Institutions." *Tribal College Journal of American Indian Higher Education*, 16(3), (Spring), 10–14.

Brayboy, B. M. J. (2004). "Hiding in the Ivy: American Indian Students and Visibility in Elite Educational Settings." *Harvard Educational Review*, 74(2), 125–152.

Cajete, G. (1994). *Look to the Mountain, An Ecology of Indigenous Education*. Skyland, NC: Kivaki Press.

Child, B. (1995). *Boarding School Seasons*. Lincoln, London: University of Nebraska Press.

Clark, R. O. (1972). "Higher Education Programs for American Indians" [Electronic version]. *Journal of American Indian Education*, 12(1). Retrieved May 7, 2006, from http://jaie.asu.edu/v12/V12S1hig2.html.

Cummins, J. (2003). "Challenging the Construction of Difference as Deficit: Where are Identity, Intellect, Imagination, and Power in the New Regime of Truth?" In P. P. Trifonas (ed.), *Pedagogies of Difference, Rethinking Education for Social Change*. New York, London: Routledge Falmer. 206–219.

Darling-Hammond, L. (1999). "Educating the Academy's Greatest Failure or Its Most Important Future?" *Academe*, 85(1), 26–33.

Deloria, V. Jr. and Wildcat, D. R. (2001). *Power and Place, Indian Education in America*. Golden, CO: Fulcrum Resources.

Delpit, L. (1988). "The Silenced Dialogue: Power and Pedagogy in Educating Other People's Children." *Massachusetts: Harvard Educational Review*, 58, 280–298.

Demmert, W. G. (2001). *Improving Academic Performance among Native American Students: A Review of the Research Literature* [Electronic version]. Charleston, WV: ERIC Clearinghouse on Rural Education and Small Schools.

Duran, E. and Duran, B. (1995). *Native American Postcolonial Psychology.* Albany: State University of New York.

Foster, M. (1994). "Effective Black teachers: A Literature Review." In E. Hollins, J. King, and W. Hayman (eds.), *Teaching Diverse Learners: Formulating A Knowledge Base for Teaching Diverse Populations.* Albany: State University of New York Press. 207–225.

Francis, N. and Reyhner, J. (2002). *Language and Literacy Teaching for Indigenous Education, A Bilingual Approach.* Clevedon, Buffalo, Toronto, Sydney: Multilingual Matters LTD.

Gajar, A. (1985). American Indian Personnel Preparation in Special Education. *Journal of American Indian Education*, 24, 7–15.

Gay, G., Dingus, J. E., and Jackson, C. W. (2003). "The Presence and Performance of Teachers of Color in the Profession." White paper. Washington DC: Community Teachers Institute.

George, M. L. (2005). "The Promise of Indigenous Education: A Case Study of Navajo Bilingual-Bicultural Teachers." Unpublished doctoral dissertation, University of Kansas.

Hollins, E. and Guzman, M. T. (2005). "Research on Preparing Teachers for Diverse Populations." In M. Cochran-Smith and K.M. Zeichner (eds.), *Studying Teacher Education, The Report of the AERA Panel on Research and Teacher Education.* Mahwah, NJ: Lawrence Erlbaum Associates, Inc. 477–548.

Kawagley, A. O. (1995). *A Yupiaq Worldview: A Pathway to Ecology and Spirit.* Long Grove, Ill: Waveland Press, Inc.

Ladson-Billings, G. (2000). "Fighting for Our Lives: Preparing Teachers to Teach African American Students." *Journal of Teacher Education*, 51(3), 206–214.

Loewen, J. W. (1995). *Lies My Teacher Told Me, Everything Your American History Textbook Got Wrong.* New York: Touchstone.

Manuelito, K. D. (2003). "Building a Native Teaching Force: Important Considerations." *ERIC Digest.* (ERIC Identifier: ED482324)

National Commission on Teaching and America's Future (2004). "High Quality Teacher Preparation: Higher Education's Crucial Role." Summary Report.

Ng, R. (2003). "Toward An Integrative Approach to Equity in Education." In P. P. Trifonas (ed.). *Pedagogies of Difference, Rethinking Education for Social Change.* New York, London: Routledge Falmer. 206–219.

Reyhner, J. (1992). *Teaching American Indian Students.* Norman: University of Oklahoma Press.

Reyhner, J. and Eder, J. (2004). *American Indian Education, A History.* Norman: University of Oklahoma Press.

Smith, L. T. (1999). *Decolonizing Methodologies, Research and Indigenous Peoples.* London and New York: Zed Books Ltd.

Swisher, K. G. (1997). "Preface" [Electronic version]. *Journal of American Indian Education*, 37(1). Retrieved July 16, 2006, from http://jaie.asu.edu/v37/V37S1pre.htm.

Swisher K. G.and Tippeconnic III, J. W. (eds.) (1999). *Next Steps: Research and Practice to Advance Indian Education.* Charleston, WV: ERIC Clearinghouse on Rural Education and Small Schools. (ERIC Document Reproduction Service No. ED427902).

Tharp, R. G. (2006). "Four Hundred Years of Evidence: Culture, Pedagogy, and Native America." *Journal of American Indian Education*, 45(2), 6–25. Special issue.

Thornton, R. (1998). "The Demography of Colonialism and 'Old' and 'New' Native Americans." In R. Thornton (ed.), *Studying Native America, Problems and Prospects.* Madison: University of Wisconsin Press. 17–39.

Tierney, W. G. (1991). "Native Voices in Academe." *Change*, 23(2), 36–44.

Tippeconnic, J. W. III. (1983) "Training Teachers of American Indian Students." *Peabody Journal of Education*, 61(1), 6–15.

U.S. Census Bureau. (2001a). *Profiles of General Demographic Characteristics 2000.* (Table DP-1. Profile of General Demographic Characteristics, 2000). Washington, DC.

———. (2001b). *Census 2000 Summary File 1.* (Table 4. Population by Race Alone or in Combination and Age for the United States, 2000). Washington, DC: With Author.

U. S. Department of Education. (1991). *Indian Nations At Risk: An Educational Strategy For Action.* Washington, DC: U.S. Department of Education.

Zumwalt, K. and Craig, E. (2005). "Teachers' Characteristics: Research on the Demographic Profile." In M. Cochran-Smith and K. M. Zeichner (eds.), *Studying Teacher Education, The Report of the AERA Panel on Research and Teacher Education.* Mahwah, NJ: Lawrence Erlbaum Associates, Inc. 111–156.

CHAPTER 14

Reframing Refugee Education in Kenya as an Inclusionary Practice of Pedagogy

Wangari Pauline Gichiru and Douglas B. Larkin

> *There is no neutral education. Education is either for domestication or for freedom.*
>
> —*João da Viega Coutinho (quoted in Freire, 1970, p. vi)*

There can be no doubt that large-scale humanitarian assistance is a necessary element of today's global geopolitical environment. Wars, civil strife, environmental catastrophes, and other events compel significant numbers of people into situations in which they face a dire need for food, shelter, security, and a path toward a return to lives interrupted by tragedy. This assistance is currently provided in a number of ways through governmental, quasi-governmental, and nongovernmental organizations.[1] Over the past decade, these actors have also come to view education as an equally important part of the humanitarian assistance they provide. While education was earlier seen by donors and planners as primarily a development activity and excluded as a form of humanitarian assistance, it is now considered to be an essential component for meeting the immediate needs of refugees (Sinclair, 2001; Sommers, 2001; UNESCO, 2000a). The debates over how education ought to occur in emergencies and after they are over are framed in particular and pragmatic ways, and they necessarily focus on how to go about providing education to refugees in the most effective manner.

Yet the questions regarding both the means and the ends of such education ultimately entail some inquiry into the character of the pedagogies enacted with refugee children. Our concern here is that such pedagogies and associated policies do not further exclude the already marginalized people. Rather, meaningful education in these circumstances must necessarily give the people the tools to name and act upon their own circumstances. Such education allows for a form of assimilation into the wider society that is non-oppressive and yet inclusive. It enables civic participation, provides a vision for the future, and develops and realizes each individual's potential for growth as a human being.

In this chapter, we argue that the issues of refugee education cannot be disentangled from the context of education that exists in the host country. Toward this end, we will examine the case of refugee education in Kenya within the context of Kenya's educational system and governmental policies, and then present an alternate view of how refugee education might be constructed with pedagogies of inclusion in mind.

What Does It Mean to Be a Refugee?

According to Article 1.2 of the United Nations' 1951 convention relating to the status of refugees, the term applies to any person who "owing to a well founded fear of being persecuted for reasons of race, religion, nationality, membership of a particular social group or political opinion, is outside the country of his nationality and is unable to or, owing to such fear, is unwilling to avail himself to the protection of that country" (UNHCR, 1951).

The modern era of massive numbers of refugees began shortly after World War II, when approximately one million people were uprooted and displaced as a result of the war. The question of how to best deal with the massive humanitarian needs of this population led to the creation in 1950 of the Office of the United Nations High Commissioner for Refugees (UNHCR) by the newly formed United Nations. Despite the fact that World War II affected many areas around the globe, it is interesting to note that this policy document referred explicitly only to postwar Europe. The definition of a refugee was expanded by the convention's 1967 protocol, and later by regional conventions, to include persons fleeing war or other violence in places other than Europe.

Though who counts as a refugee varies across world regions, one common factor is that to be recognized as a refugee there has to be an international border crossing. In this chapter, we refer to the country that first receives refugees across their border as the *host country*. Upon crossing the border, refugees are often obliged to trade "elements of citizenship in their own country for safety on terms decided by host governments" (Smith, 2004). These elements of citizenship may include freedom of travel, property and suffrage rights, and the protection by their home government, among other things.

In order to provide immediate emergency assistance to large numbers of fleeing people who may be exhausted, frightened, and hungry, setting up refugee camps is often a logistical necessity. In these safe havens, host governments and/or aid organizations provide lifesaving provisions including food, medicine, and shelter. While camps are intended to provide a temporary solution for refuge-seeking people, they have become increasingly more permanent because of the protracted nature of conflicts, with the current average stay for two-thirds of global refugees being seventeen years, in contrast to an average stay of nine years in the early 1990s. In this chapter, our primary focus will be on refugee education as it exists in these refugee camps.

EDUCATION SYSTEMS IN COUNTRIES THAT HOST REFUGEES

Over the past decade, a number of thorough literature reviews have appeared that sift through and summarize what is known about the field of refugee education, specifically as it occurs in emergency situations (e.g., Burde, 2005; Seitz, 2004; Sinclair, 2001). A large body of work also describes educational programs in various refugee camps around the world (Dryden-Peterson, 2003; Jackson, 2000; Moro, 2002) as well as various approaches to refugee education, such as peace education, HIV/AIDS education, and landmine awareness education (Lyby, 2001; Sommers, 2001). It is telling that none of the literature that were reviewed connected refugee education in camps with the condition of education in the host country in general,[2] and it also does not discuss the nature of the pedagogy employed in schools located in refugee camps—particularly as it related to social exclusion.

Questions over what is to be the purpose and nature of education in refugee camps remain closely tied to the education students might otherwise have received if they had not been forced to flee from their country of origin. The majority of educational systems in sub-Saharan Africa are themselves based upon colonial models of schooling and retain their socialization features—such as the reproduction of social inequality, pedagogical models that confer expertise to authority, and the development of a national identity—which are the artifacts of colonization. It is likely that many of the conditions of schooling in these systems are also found within schools in refugee camps that are located within the host country.

THE GLOBAL CONTEXT FOR REFUGEE EDUCATION IN KENYA

At the reaffirmation of the vision set out over a decade ago at the World education Forum in Dakar, Senegal, 180 countries pledged to ensure that, "by 2015, all children and particularly girls, children in difficult circumstances (including those affected by war) and those belonging to ethnic minorities, will have access to free and compulsory primary education of good quality" (UNESCO, 2000a, p. 15). This is one among many other instruments to which Kenya has pledged in order to reiterate its commitment to education as a basic human right.[3] From these and other international agreements, one might conclude that structures have been put in place to ensure that refugees in Kenya not only become full participants in the Kenyan educational system but would also enjoy the freedom to use that education and be able to contribute to society in meaningful ways.

Kenya's history of relative peace and political stability in East Africa might support this view. However, the nature of the ethnic violence following the 2007 general election calls into question the capacity of Kenyan society to allow refugees to integrate and become full participants and contributors in society. Furthermore, a glance at Kenya's current economic situation might make such optimism seem unrealistic. The government continually struggles with population pressures, widespread unemployment, and a staggering external debt that constrains domestic spending.[4] With only 7 percent of its land arable, Kenya has on an almost annual basis dealt with recurring drought and occasional flooding during the rainy season. As a regional hub for trade and finance in East Africa, Kenya has also been hampered by corruption and by reliance on marketing primary goods whose international prices have remained low.

In the light of these and other issues, the international promise of Kenya as a "safe haven" that offers full participation in civic life seems much more difficult to achieve in reality than on paper. Despite these conditions, Kenya is still one of the most coveted refugee destinations in the region, because these concerns often pale in comparison to the troubles that compel refugees to flee their home countries in the first place.

Until 1991, the government of Kenya determined the legal status of newly arrived refugees on a case-by-case basis. However, the arrival that year of some 400,000 refugees in Kenya, among whom were the famous "lost boys of the Sudan" (Bixler, 2005), led to the collapse of the system of individual status determination. Verdirame (1999) describes the effects of this influx of refugees in detail. He notes that these large numbers simply exceeded Kenya's capacity to absorb them through the originally generous, though somewhat laissez-faire, refugee policy. Where formerly the refugees had few obstacles to local integration and the enjoyment of basic rights including education, once UNHCR took over this role, the government of Kenya adopted the stance that the refugees were "UNHCR's problem" (Verdirame, 1999). Subsequently, Kenya became a transit country, where refugees were allowed to remain only in the camps until a durable solution to their case was found.

Other issues also have influenced refugee education in Kenya. With the bombing of the U.S. Embassy in Nairobi in August 1998 and the general increase in terrorist threats worldwide after the events of September 11, 2001, the Government of Kenya has toughened its stance on refugees—who are increasingly being seen as a security threat. Also, apart from the murky issues of legal status that hinder full refugee participation in education, the locations of the camps themselves historically is a reason to have low enrollments, because they are in arid or semiarid lands inhabited by nomadic communities (Ministry of Education Science and Technology, 2004).

Clearly, it has been difficult for Kenya to live up to its international agreements. As of 2008, a national Refugee Bill had still not been enacted, and as a result, Kenya currently lacks a "legal framework for asylum seekers and refugees, who continue to be treated in accordance with the aliens Act" (UNHCR, 2005, p. 188). Factors that cause this include the increasing influx of refugees that continue to overwhelm the country's capacity to serve them, Kenya's crippling external debt, and weaknesses in enforcement and accountability mechanisms in international law. It is also salient to point out that, like in many countries, refugee issues are apt to become politicized in Kenya and, as such, are subject to domestic policy decisions and debates.

While the government's stand is that the main aim of education is to earn a living and to be able to fit into a social world (Ministry of Education, Science and Technology, 2004), for more than a decade the majority of the refugees in Kenya have been required to reside in camps. This contradiction, namely, that refugee education in Kenya seems to prepare students for a society in which they are subsequently denied participation, is problematic to say the least. It calls into question the rationale for education in its most basic sense, and the argument that education is either for domestication or for freedom takes on a tangible quality that must be faced squarely.

REFUGEE EDUCATION WITHIN THE CONTEXT OF KENYA'S EDUCATION SYSTEM

To further understand the context in which refugee education exists in Kenya, it is important to look at a brief history and current conditions of education there. In her book *Growing Up in Kenya: Rural schooling and Girls*, Mungai (2002) explains that the foundation of modern education in Kenya was laid by missionaries in the 1800s in an effort to spread Christianity. They later taught practical subjects such as agriculture and training, and until the British colonial government's review of education in 1949 the main aim of education was not to teach academic skills but rather to impart basic skills like carpentry and masonry to increase production on settlers' farms. At independence in 1963, the new government set ambitious goals of providing an education to train Kenyans for work in modern sectors of the economy as well as to cater for the needs of the still predominantly rural economy. Despite these goals, Mungai (2002) states that education today is generally urban oriented and largely European in its assumptions and methods.

As a nation where approximately forty different languages and dialects are spoken, fostering national unity has been a crucial goal in Kenyan education. To do this, Kenya has used its national languages of Kiswahili and English for instruction, with local languages occasionally being used alongside Kiswahili and English in the early primary grades. The cultural and linguistic heterogeneity of the refugees that come to Kenya further complicates the language situation in schools.

The concept of *Harambee* (Kenyatta, 1965) as a national development philosophy is too important to be overlooked in this context. Articulated numerous times by Jomo Kenyatta, the first president of Kenya, this idea based on traditional notions of collective social action on a local scale was transformed into a national development strategy.[5] The literal meaning of harambee—"pull together"—conveyed a sense of national ownership over different development activities, among which was the expansion of existing schools and construction of new ones to benefit all. This was especially important as Kenya's population grew dramatically after independence. Mungai

(2002) states that the number of school going children increased from approximately 1 million pupils in 1964 to almost 5 million pupils in 1986. This was a direct consequence of population growth combined with greater access to formal education, fostered by the initially free primary education.

In 1985, a series of both structural and curricular reforms was introduced in order to better meet the needs of Kenyans who did not progress beyond the primary and/or secondary levels. These reforms included greater opportunity to develop vocational skills, as well as a curriculum more centered within a Kenyan context. Though schooling was still nominally free, in the spirit of Harambee, parents still had to contribute to "building funds," to buy uniforms, textbooks, and to pay for school building expansions (Mungai, 2002).

In 2003, the newly elected government embarked on an ambitious project to revitalize the education sector by the implementation of major reforms including the long-awaited introduction of free primary education (Ministry of Education Science and Technology, 2004). One immediate consequence of offering free primary education was the overcrowding of schools, a situation that continues to place an enormous strain on the teachers and to a larger extent on the Government.

Beyond Pedagogies of Exclusion in Refugee Education in Kenya

It seems a bit distant perhaps, writing from the United States, to attempt to influence the classroom practices of the teachers of refugees (who, it must be noted, may be refugees themselves) in the potentially volatile environment of refugee camps. Such teaching work is carried out in conditions that might try the patience, strength, and skills of the most accomplished teachers from anywhere in the globe. What can be said here that could conceivably affect what occurs in these classrooms that are isolated from, yet subject to, the educational systems and policies of the host countries in which they are located?

Rather than present a prescriptive series of recommendations for such teachers to follow, we present below some themes drawn from the educational research literature on effective teaching in culturally diverse societies (Cochran-Smith, Davis, and Fries, 2004; Grant, 1997; Grant and Secada, 1990; Ladson-Billings, 1994, 1999, 2001; Villegas, 2008; Villegas and Lucas, 2002; Zeichner, 1996). While much of this research differs significantly in terms of the local school contexts from that of Kenyan camps, we feel strongly that it has the potential to aid decision-making in refugee education, both by refugee education policy-makers and by teachers themselves. In the previous sections, we have detailed a number of issues relating to the structure of educational experiences in Kenya generally and in its refugee camps. Here, we wish to refer to specific pedagogies that can be employed with refugee students in schools located in refugee camps.

In a wide-ranging review of the literature on effective teaching of students in a culturally diverse society, Zeichner (1996) has compiled a list of findings that would seem to apply equally well to refugee education in Kenya as they would to diverse classrooms in the United States. These include steps to ensure the following:

- Teachers have a clear sense of their own ethnic and cultural identities.
- High expectations for the success of all students (and a belief that all students can succeed) are communicated to all students.
- Teachers are personally committed to achieving equity for all students and believe that they are capable of making a difference in their students' learning.
- Teachers have developed a bond with their students and cease seeing their students as "the other."
- Students are provided with an academically challenging curriculum that includes attention to the development of higher level cognitive skills.
- Instruction focuses on the creation of meaning about content by students in an interactive and collaborative learning environment.
- Learning tasks are often seen as meaningful by students.
- The curriculum is inclusive of the contributions and perspectives of the different ethnocultural groups that make up the society.
- Scaffolding is provided by teachers that links the academically challenging and inclusive curriculum to the cultural resources that students bring to school.
- Teachers explicitly teach students the culture of the school and seek to maintain students' sense of ethnocultural pride and identity.
- Parents and community members are encouraged to become involved in students' education and are given a significant voice in making important school decisions in relation to program, that is, sources and staffing.
- Teachers are involved in political struggles outside of the classroom aimed at achieving a more just and humane society (149).

Drawing upon this work as well as the larger body of scholarship in U.S. teacher education, we take three main themes and describe their implications for the teachers of refugee students in Kenya. These three themes are: (1) the beliefs of teachers, (2) the selection and enactment of curriculum, and (3) the actions of teachers in classrooms.

The Beliefs of Teachers

Teachers need to have high expectations for their students, but these high expectations ought not to be limited to academic achievement as measured by Kenyan graduation exams. Teachers who view their students as passive recipients of delivered knowledge deny them the opportunities to construct and name their own realities, while those who believe that the lived experiences of their students are valuable resources are more likely to tap into them to teach subject matter.

Pedagogies of exclusion are sustained when teachers believe in the superiority of models of teaching introduced during the period of colonialism, which emphasize an individualistic approach to education that denies the cultural resources of students and views schools as mechanisms for sorting students by intellectual merit rather than the development of human potential. To move beyond these pedagogies, teachers must begin to view their students as full of promise, with the goal of helping them develop the academic, cultural, and social tools they will need to act on and improve their life situations.

The Selection and Enactment of Curriculum

The recent global expansion of peace education, conflict resolution, and landmine awareness curricula through the refugee education world (Sinclair, 2002) is a promising development, because these are centered on students' present and future lives, and encourage them to draw upon their experiences to plan for future action. Yet these are not the only areas of curriculum in which this is possible. Drawing from the work of Moses and Cobb (2001) who developed an algebra curriculum based on the lived experiences of their students in urban environment, a teacher of refugee students might connect the content of mathematics to the economics and social dynamics of the camps.

Freire (1970) reminds us that the purpose of literacy is to read and write the world as a text. A teacher who teaches science by examining the water and electrical systems of the camps, and then helps them connect this knowledge to broader resource access issues in Kenya, is engaging in a much more powerful form of teaching than a teacher who writes textbook notes on the board for students to copy and memorize. Teachers who actively research their students' lives for the purposes of identifying the strengths they bring to learning enact a pedagogy that is necessarily inclusive by valuing students' home cultures (González, Moll, and Amanti, 2005).

The Actions of Teachers in Classrooms

Teachers who enact an ethic of caring with their students, both in regard to their academic work and their socioemotional growth, perform differently than teachers who see their work as simply delivering curriculum. Approaches to the portrayal of content, classroom management, grouping strategies, conflict resolution, and the assessment of student learning can work to develop students' critical consciousness and reinforce their own cultural competence while helping them learn the subject matter.

It is not uncommon in East African classrooms (in and out of the camps) to find that students must share limited physical resources, such as in the case when two, three, or even more students must share a single desk. What is less common is to see students sharing intellectual resources. Effective teachers recognize that students can work together to produce knowledge and connect it to more common constructions of academic subjects. The Harambee spirit need not end at the classroom door. Such teachers regularly access the language and experiences of their students when interacting with students and their families. Even for teachers unfamiliar with the languages of their students, beginning to study basic phrases (especially numbers for mathematics) in the languages, even multiple ones, may be no more difficult than learning student names, and this sends a message to the students that their home cultures will be valued as they become accustomed to a new one.

THE CHALLENGE AND PROMISE OF INCORPORATING REFUGEES INTO THE KENYAN EDUCATION SYSTEM

As our discussion has shown, refugee Education in Kenya is run by agencies that have minimum interaction with government-run schools, because by virtue of the isolation of refugee camps from Kenyan civil society they are outside the country's curriculum, regulations and administrative set up . While Kenya's Ministry of Education, Science and Technology has minimal responsibility for education in the Kenyan camps, Kenya still needs effective policies that address the needs of the many refugees who reside outside the camps in several parts of the country in small clusters. It is likely that these refugees—who have unique needs and are scattered in several urban regions in Kenya—may cause a significant extra burden on the already hard-pressed education system. In overcrowded schools with an ever-increasing number of refugee students who may not be receiving specialized attention, teachers may simply be oblivious to the unique needs of these students.

Over the years, both the Government of Kenya and the agencies providing education in the camps have developed a vast base of knowledge and expertise. With the availability of other international networks, such as the International Network for Education in Emergencies, sharing of resources and insights has become increasingly easier. We argue, however, that this knowledge base, while important in its own right, should be used for more than just furthering the status quo of providing long-term education in camps. If incorporation of refugee students (not to mention internally displaced persons) into the host country educational systems as equals is seen as the goal, then this knowledge base will be a crucial resource for teachers and education policy-makers.

By 2005, Kenya was hosting roughly 251,000 refugees from nine neighboring nations,[6] who were consolidated into three locations: the Dadaab camp along the Somalia border, the Kakuma camp near Sudan, and Nairobi. These camps currently receive refugees at a rate of 2,000 to 3,000 new arrivals per month (Allen, 2006), many of whom arrive by foot after perilous journeys and take up residence in the camps in makeshift tents made of branches and cloth. Even though some individuals and families escape the camps and successfully integrate themselves in the larger Kenyan society, the majority of refugees will spend a long time in these camps. Many of these refugees, though stationed in Kenya, are not really in Kenya because the gates through which they might gain wider access to Kenyan society are closed to them.

The presence of camps in Kenya is not a recent phenomenon, but it is one that ought to be troubling for its historical echoes of colonialism (Elkins, 2005). In the 1950s, during the period of Emergency, the British colonial government, in an attempt to repress uprisings from local independence movements, placed large numbers of people in detention camps and *ishaggi* or forced settlements. While one might question the historical parallels between the forced internment of ethnic groups during the colonial era and current mass movements of refugees into Kenya, the setting of the "problem," namely, "How do we educate those who must remain indefinitely within camp borders?" is indeed similar.

In another historical echo, owing to global increase in migratory flows that have also come with terrorism threats, Kenya has followed the Western trend of being cautious about selectively determining entry to the country. Lately, many countries (including the United States, from which we write) have been extremely selective in the way they admit refugees, as shifting national policies continue to privilege the admittance of some refugees over others. Those admitted are usually ones who are perceived to be able to contribute economically, or whose entry is consistent with the political interests of the admitting country's government (Zucker and Zucker, 1987).

Even though a UNESCO report mentions additional grants being given to cater for the "hard to reach groups" (UNESCO, 2000b, 3), the reality of the situation is that this has not done much to increase the school enrollment levels of refugees. Today, even though a few students have been offered spots in schools within the surrounding community, it is reasonable to ask what impact these grants have had on the refugee community as a whole. Enrollment levels are still very low, and post-schooling opportunities to use their education (especially to better their own lives) are scarce. In Dadaab camp, for example, out of 45,700 school-age children, 66 percent are enrolled in primary schools but only 25 percent of this population progresses to secondary schools (UNHCR, 2005). This is lower than the enrollment rates for much of the country, but significantly higher than those in North Eastern province where the camp is located (Saitoti, 2004).

The challenge and promise of incorporating refugees into the fabric of Kenya's society depends on both giving them the necessary tools, particularly dominant languages like Kiswahili and English, as well as the opportunity to apply their education in the job market. Giving students physical access to host country schools and ensuring these refugees are not subject to discrimination within the educational system would seem an important first step. Additionally, teachers should be supported in their efforts to employ pedagogies that recognize and value the inherent worth of the cultures and languages the refugees bring with them from their home countries. Unfortunately, Kenya's system of education currently does little for refugees in this regard. Many teachers are not equipped with the necessary skills to reach out to these groups of people whose numbers continue to increase and who are not likely to leave the country in the near future.

It would be worthwhile to ask, how do we expect anything different if we deny the refugees a chance to use their own talents and education toward nation building? What would be gained if the refugees were allowed to join schools and blend into the society of their host country? Is Kenya really as "full" as it claims or is it just a political posture put in place to deny refugees their right to an asylum and an education?

It is hopeful to note that the goal of acculturating refugees may work in a positive way on the Kenyan educational system if one of the approaches serves to give greater value to the resources and strengths that all students bring to their education. The aim of creating an educated population that is able to connect to its past through language and culture and still become productive participants in national economic and political practices is a goal that would benefit all. This process would shift the norms away from a paradigm of exclusion—a situation that serves to divide people and pit them against one another in a battle for scarce resources—to one in which the goal is specifically one that seeks to include all in a common struggle. This would be the same sort of common struggle that is linked culturally and historically to the national spirit of *harambee* in Kenya.

NOTES

1. Examples of governmental organizations include the United States Agency for International Development and the German governmental organization Gesellschaft für Technische Zusammenarbeit (GTZ). Quasi-governmental organizations include the United Nations and its subgroups such as the UN High Commissioner of Refugees, and nongovernmental organizations include Save the Children, Jesuit Relief Services, and many others.
2. With the exception of the concern that it should not be superior to that of the host country, for fear of creating an incentive for refugees to remain in the camps as opposed to being repatriated (Sinclair, 2001).
3. Including 1948 Universal declaration of Human rights and the Children's Act of 2001.
4. Kenya has a population of over 32 million people, 40 percent of whom are unemployed, and an external debt of $US6 Billion (Kenya High Commission—United Kingdom, 2006).
5. By including Harambee as an organizing philosophy here, we are drawing upon Ladson-Billings' (2000) notion of "ethnic epistemologies" to emphasize a system of knowing outside the Eurocentric paradigm that has action for the collective good at its core.

6. The refugees are mainly from Somalia, Sudan, and Ethiopia and to a smaller extent from Burundi, DRC, Eritrea, Rwanda, and Uganda.

References

Allen, K. (August 19, 2006). "Getting Back to Business in Somalia." *BBC News, Kenya.*

Bixler, M. (2005). *The Lost Boys of Sudan: An American Story of the Refugee Experience* Athens, GA: University of Georgia Press.

Burde, D. (2005). *Education in Crisis Situations: Mapping the Field.* Washington, DC: Creative Associates/USAID.

Cochran-Smith, M., Davis, D., and Fries, K. (2004). "Multicultural Teacher Education: Research, Practice, and Policy." In J. A. Banks and C. A. M. Banks (eds.), *Handbook of Research on Multicultural Education* (2nd ed.). San Francisco: Jossey-Bass. xxvi, 1089.

Dryden-Peterson, S. (2003). *Education of Refugees in Uganda: Relationships Between Setting and Access* (No. RLP Working Paper No. 9). Kampala, Uganda: Refugee Law Project.

Elkins, C. (2005). *Imperial Reckoning: The Untold Story of Britain's Gulag in Kenya* (1st ed.). New York: Henry Holt.

Freire, P. (1970). *Pedagogy of the Oppressed.* New York: Herder and Herder.

González, N., Moll, L. C., and Amanti, C. (2005). *Funds of Knowledge: Theorizing Practices in Households, Communities, and Classrooms.* Mahwah, NJ: L. Erlbaum Associates.

Grant, C. A. (1997). "Critical Knowledge, Skills, and Experiences for the Instruction of Culturally Diverse Students: A Perspective for the Preparation of Preservice Teachers." In J. J. Irvine (ed.), *Critical Knowledge for Diverse Teachers & Learners.* Washington, DC: American Association of Colleges for Teacher Education. 223.

Grant, C. A. and Secada, W. (1990). "Preparing Teachers for Diversity." In W. R. Houston, M. Haberman, and J. P. Sikula (eds.), *Handbook of Research on Teacher Education: A Project of the Association of Teacher Educators.* New York: Collier Macmillan. 403–422.

Jackson, T. (2000). *Equal Access to Education: A Peace Imperative for Burundi.* United Kingdom: International Alert.

Kenya High Commission–United Kingdom. (2006). *General Information on Kenya.* Retrieved November 26, 2006, from http://kenyahighcommission.net/kenyageninfo.html.

Kenyatta, J. (1965). *Harambee! The Prime Minister of Kenya's Speeches, 1963–1964, from the Attainment of Internal Self-Government to the Threshold of the Kenya Republic.* Nairobi and New York: Oxford University Press.

Ladson-Billings, G. (1994). *The Dreamkeepers: Successful Teachers of African American children* (1st ed.). San Francisco: Jossey-Bass Publishers.

———. (1999). "Preparing Teachers for Diverse Student Populations: A Critical Race Theory Perspective." *Review of Research in Education*, 24, 211–247.

———. (2000). "Racialized Discourses and Ethnic Epistemologies." In N. Denzin and Y. Lincoln (eds.), *Handbook of Qualitative Research* (2nd ed.). Thousand Oaks, CA: Sage.

———. (2001). *Crossing over to Canaan: The Journey of New Teachers in Diverse Classrooms* (1st ed.). San Francisco: Jossey-Bass.

Lyby, E. (2001). "Vocational Training for Refugees: A Case Study from Tanzania." In J. Crisp, C. Talbot, and D. B. Cipollone (eds.), *Learning for a Future: Refugee Education in Developing Countries*. Geneva, Switzerland: UN High Commissioner for Refugees. 217–259.

Ministry of Education Science and Technology. (2004). *Development of Education in Kenya*. Retrieved August 1, 2006, from http://www.ibe. unesco.org/International/ICE47/English/Natreps/reports/kenya. pdf#search=%22Ministry%20of%20Education%20Science%20and%20 Technology%2C%20(2004).%20Development%20of%20Education%20 in%20Kenya%22.

Moro, L. N. (2002). *Refugee Education in a Changing Global Climate: The Case of Sudanese in Egypt*. 46th Annual Meeting of Comparative and International Education Society: Orlando, FL.

Moses, R. P. and Cobb, C. E. (2001). *Radical Equations: Math Literacy and Civil Rights*. Boston: Beacon Press.

Mungai, A. (2002). *Growing up in Kenya: Rural Schooling and Girls*. New York: Peter Lang.

Saitoti, G. (2004). *Education in Kenya: Challenges and Policy Responses*. Retrieved July 22, 2008, from http://www.cfr.org/content/meetings/ CUE%20Meetings/CFR_Saitoti_Presentation_April_2004.ppt

Seitz, K. (2004). *Education and Conflict: The Role of Education in the Creation, Prevention and Resolution of Societal Crises—Consequences for Development Cooperation*. Rossdorf, Germany: Deutsche Gesellschaft für Technische Zusammenarbeit (GTZ) GmbH—German Technical Cooperation.

Sinclair, M. (2001). "Education in Emergencies." In J. Crisp, C. Talbot, and D. B. Cipollone (eds.), *Learning for a Future: Refugee Education in Developing Countries*. Geneva, Switzerland: UN High Commissioner for Refugees. 1–84.

———. (2002). "Planning Education in and after Emergencies. Fundamentals of Educational Planning." *United Nations Educational, Scientific, and Cultural Organization, International Inst. for Educational Planning*, 7: Paris.

Smith, M. (2004). *Warehousing Refugees: A Denial of Rights, a Waste of Humanity*. Washington: US Committee for Refugees.

Sommers, M. (2001). "Peace Education and Refugee Youth." In J. Crisp, C. Talbot, and D. B. Cipollone (eds.), *Learning for a Future: Refugee Education in Developing Countries*. Geneva, Switzerland: UN High Commissioner for Refugees. 163–216.

UNESCO. (2000a). *Dakar Framework of Action. World Education Forum*. Retrieved December 2, 2005, from http://unesdoc.unesco.org/images/

0012/001211/121147e.pdf#search=%22world%20education%20forum%20in%20dakar%22

UNESCO. (2000b). *The EFA 2000 Assessment Country Report: Kenya.* Retrieved August 26, 2006, from http://www2.unesco.org/wef/country reports/home.html

UNHCR. (1951). *Convention relating to the status of refugees.* Retrieved November 17, 2006, from http://www.unhchr.ch/html/menu3/b/o_c_ref.htm

———. (2005). *UNHCR Global Report 2005.* Retrieved November 25, 2006, from http://www.unhcr.org/cgi-bin/texis/vtx/template?page=publ&src=static/gr2005/gr2005toc.htm

Verdirame, G. (1999). "Human Rights and Refugees: The Case of Kenya." *Journal of Refugee Studies,* 12(1), 55–77.

Villegas, A. M. (2008). "Diversity and Teacher Education." In M. Cochran-Smith, S. Feiman-Nemser, D. J. McIntyre, and Association of Teacher Educators (eds.), *Handbook of Research on Teacher Education: Enduring Questions in Changing Contexts* (3rd ed.). New York: Routledge; Co-published by the Association of Teacher Educators. 551–558.

Villegas, A. M. and Lucas, T. (2002). *Educating Culturally Responsive Teachers: A Coherent Approach.* Albany: State University of New York Press.

Zeichner, K. (1996). "Educating Teachers for Cultural Diversity." In K. M. Zeichner, S. L. Melnick, and M. L. Gomez (eds.), *Currents of Reform in Preservice Teacher Education.* New York: Teachers College Press. 133–175.

Zucker, N. L. and Zucker, N. F. (1987). *The Guarded Gate: The Reality of American Refugee Policy.* Harcourt Brace Jovanovich: San Diego. Retrieved from http://www.unhcr.org/partners/PARTNERS/4381c5832.pdf.

Epilogue: On to Even More Engaging and Challenging Transnational Conversations about Pedagogies of Inclusion

Carl A. Grant

Most definitions of an epilogue are fairly consistent: "An epilogue, or epilog, is a piece of writing at the end of a work of literature or drama, usually used to bring closure to the work" (en.wikipedia.org/wiki/Epilogue); or, "Additional text at the end of the book, that provides readers with additional information on the subject" (aalbc.com/writers/publishing_glossary.htm). I prefer this second definition because the statement, "provides readers with additional information on the subject," better fits with my intentions in writing this epilogue, as you will soon see.

Readers of this volume will discover a rich fund of knowledge, skills, and dispositions about pedagogies beyond exclusion, including insights about educational policy and practice from an international perspective. Because of the informative nature of this collection, it may be prudent to review the chapters that best connect with your teaching and research responsibilities. And if, in carrying out this task, you are pleased with the ideas you learn about pedagogies beyond exclusion and you act on these ideas, the editors' goal for this book—to have conversations that lead to action about pedagogies beyond exclusion—is achieved.

Taking off my editor's hat (it took four years for this project!) and reading the chapters to inform my own teaching and scholarship, I come away with a deeper understanding of and appreciation for pedagogy. Equally significant, however, I come away with a few questions and/or arguments. The most immediate and critical question/argument has to do with the definition(s), conceptualization and articulation of pedagogy (pedagogies), as well as pedagogy as a form of social administration. In epilogue parlance, my question/argument is guided by that second definition: I am pushing us even further in our understanding of pedagogy by providing "additional information on the subject."

FUTURE CONVERSATIONS ABOUT PEDAGOGIES BEYOND EXCLUSION

This "additional information on the subject" comes first as a recommendation and invitation and then as a question. First, the recommendation and invitation is for chapter authors in this book—along with other scholars who teach, research, and write about pedagogies beyond exclusions—to meet once again in Thessaloniki, Greece, to continue this important conversation. Second, my question, which I will make at the opening of our conference, is: "What can we do to further refine, promote, and make clear our conceptualization of pedagogies of inclusion, and how can we pay better attention to how pedagogies are used for 'social administration?'"

My reason for this question has less to do with the scholarship in this book on pedagogies of inclusion and more to do with how pedagogy is discussed in the education literature at large. Let me explain by starting with definition(s) of pedagogy in the education literature. I do this because I am hoping to push the scholarship of exclusion even further than where we stand now.

DEFINITION(S) OF PEDAGOGY

A cursory review of the education literature suggests that a fairly common definition of pedagogy exists. However, it should be noted that many education texts written for undergraduates and graduates over the past four decades do not include pedagogy in the index. Rather, pedagogy became significant in the teacher education discourse with the publication of Paulo Freire's first book, *Pedagogy of the Oppressed*, in 1970, and with Lee Schulman's discussion of pedagogical content knowledge in 1987. I believe it fair to add that pedagogy as a term and idea was slow to gain traction. Here, it is important to note an exception—culturally relevant and responsive pedagogy—but also to acknowledge that an adverb (culturally) and an adjective (relevant or responsive) come before pedagogy. With the use of these modifiers, it makes for a special kind of pedagogy and does not reflect an understanding of pedagogy more generally.

The common definition of "pedagogy," without the adverb and adjective, usually contains the following identifying characteristics: teachers, learning/students, and sometimes assessment and curriculum. Many definitions of pedagogy reveal a blindness about students and very few definitions of pedagogy address, classify, or identify the students involved, rarely even mentioning their ascribed characteristics (e.g., race, socioeconomic class, gender). In addition, even fewer definitions of pedagogy pay attention to students' culture and history. Finally, definitions of pedagogy are often silent about the civic or political nature of students, teachers, assessments, and the content of curriculum and pedagogy as a form of social administration (with the exception of Popkewitz, 2001, 314; see below). The following are some examples of definitions of pedagogy that we find in the education literature:

- "I now believe pedagogy is the reciprocal, interactive, and dynamic process between teaching and learning. It is what goes on everyday in great classrooms. There is no one perfect pedagogy. Multiple paths exist on the path to good teaching and learning." *Joanwink.com* (retrieved May 12, 2008).
- "The art, practice or profession of teaching; especially, systematized learning or instruction concerning principles and methods of teaching." yahoo.com/question/index?qid=2007111708 3601AA26zE1—33k (retrieved August 15, 2008).
- "Pedagogy, the careful investigation of the processes of teaching and learning" (Bestor, 2008, 948).
- "[A] synonym for teaching…describes the rational values, the personal engagement, [and] the pedagogical climate…[It involves] problematizing the conditions appropriate [to] educational practices and aims to provide a knowledge base for professionals in their work with children [students]" (Loughran, 2008, 1178).
- "Pedagogy is a practice of social administration of the individual" (Popkewitz, 2001, 314).

Pedagogy is also sometimes referred to as the correct use of teaching strategies. Common definitions of pedagogy consider it a specific way of organizing formal education in educational settings, categorized by curriculum, instruction and assessment.

From the above definitions, undergraduate and graduate students receive little direction about considering the personhood of students, the nature and content of the curriculum, the form of assessment, or the controlling nature of pedagogies. Implicitly, pedagogy is characterized as an event/activity between teacher and learner that is absent of human conditions such as caring, heterogeneity among students, social inequities, or debates over power, privilege, and access. It is completely depersonalized and mechanized.

I believe—as we have demonstrated in this volume—that we must clearly articulate and define pedagogies that engage learners with curriculum, teaching, and assessment by explicitly taking into account their race, class, gender, religion, language, ethnicity, culture, sexuality, and (dis)ability; we must also articulate how the pedagogy we are using limits or gives access to opportunity. Also, attention should be given to the classroom climate where the pedagogy takes place, noting that in a caring and safe learning space, students are much more likely to challenge social and cultural inequities and engage in sensitive discussions about civic and social issues. Furthermore, with the influence of globalization—and here I am particularly talking about the mobility of populations that will increase student diversity in schools—our articulation of pedagogy should make clear that a "colorblind approach" to teaching students is unacceptable. Instead, pedagogies put into practice should pay careful attention to student diversity, culture, and history, as well as power relations that affect problems and issues within these areas.

CONCEPTUALIZATIONS OF PEDAGOGIES BEYOND EXCLUSION

The second aspect of my "additional information" is the question, How can we pay better attention to the ways that pedagogies are used for "social administration"? Here I want to return to Thomas Popkewitz's definition: "Pedagogy is a practice of social administration of the individual" (Popkewitz, 2001, 314). Popkewitz's definition leads me to wonder whether we should be more critical/analytical in our discussion of pedagogies beyond exclusion by more clearly advocating against the idea of pedagogy as a form of "social administration"—or at the minimum, make clear the form of social administration we are advocating and why. Popkewitz, rightly so, reminds us that in the early twentieth century, discourses on childhood, the state, and schooling discussed and advocated the development of a child/student who would become an adult who is self-disciplined, self-motivated, and a proactive participant in the current social projects of the day. Popkewitz contends that, presently, the governing principle of social administration is that the student will find his or her own solutions through problem-solving strategies.

Popkewitz states, "[A]t least since the nineteenth century, pedagogical discourses about teaching, children, and learning in schools have connected the scope and aspirations of public powers with the personal and subjective capabilities of individuals. This administration of the child embodies certain norms about the inner capabilities from which the child can become self-governing and self-reliant" (314). He goes on to argue that it is through examining the changes in the administration of the child that constructivist theories and theorists (e.g., Dewey and Vygotsky) can be understood in the past and present, and that these changes in the principles of administration of the child are not only changes in the rules of pedagogical "reason" but in the strategies by which systems of social inclusion/exclusion are constructed. In other words, pedagogy is a means by which we structure our society.

It is for this very reason—that pedagogy has been one way of structuring and passing on a system of social inclusion and exclusion—that pedagogies beyond exclusion must contain transformative qualities and have as their purpose teaching students how to govern, rather than how to be controlled. Pedagogies beyond exclusion must actively and explicitly work against the type of exclusive and inequitable social control Popkewitz describes. Giroux (2003) makes this same argument when he states:

> Our transformative pedagogies should encourage educators and students to: learn how to govern rather than be governed, while assuming the role of active and critical citizens in shaping the most basic and fundamental structures of a vibrant and inclusive democracy...Learning at its best is connected with the imperatives of social responsibility and political agency. (7–9)

Recalling the benign definitions of pedagogy above and recognizing that the social administration of pedagogy might not be currently considered in them, it is critical to challenge scholars and teachers on how they conceptualize and put into practice their pedagogies (e.g., curriculum, assessment). Pedagogies that move beyond exclusion should articulate a social justice purpose (e.g., critical pedagogy, transformative pedagogy) as opposed to a social administrative thesis, and it should support an inclusion ideology that embraces varied concerns about those who are dominated and oppressed (and denied privilege, cultural recognition, equity, etc.). In addition, pedagogies beyond exclusion should articulate the significance of intersections of overlapping categories of social identity including culture, ethnicity, race, gender, sexual orientation, class, and disability, because such social markers have traditionally served to aid the social administration of students.

Henry Giroux, Peter McLaren, bell hooks, and Paulo Freire are often associated with pedagogies beyond exclusion in that they argue for the concept of critical pedagogy. In other words, they argue for education to be transformative and emancipatory. For example, Giroux (2003) sees pedagogy as "a moral and political practice crucial to the production of capacities and skills necessary for students to both shape and participate in social life" (11). This reasoning about pedagogies moves beyond commonplace interpretations of prevailing theories, such as placing students into special classes, as well as beyond pedagogies that govern the individuals in ways that limits their life chances. Giroux (2003) states:

> Educators...should reject forms of schooling that marginalize students who are poor, black and least advantaged. This points to the necessity for developing school practices that recognize how issues related to gender, class, race and sexual orientation can be used as a resource for learning rather than being contained in schools through a systemic pattern of exclusion, punishment and failure. (10)

Paulo Freire refers to his method of teaching adults as "critical pedagogy." Freire's philosophical beliefs inform his teaching strategies in that attention is given to pupils' background knowledge, experiences, personal situations, and environment, as well as achievement goals set by the teacher and student. Such an idea about pedagogy clearly moves beyond what is too often practiced, beyond beliefs such as a deficit model—a harsh form of social administration that believes only some people can learn—and beyond commonplace interpretations and prevailing theories such as the meta-narrative of social reproduction theory—in which patterns of hierarchy, governance, abuse, and exclusion are legitimized and preserved.

In sum, I believe our future conversations need to address how we can more assertively challenge the commonplace definition of pedagogy and how we can pay better attention to the ways that pedagogies are used for social administration. With this focus, I hope that we can further develop pedagogies beyond exclusion, in other words, a social justice narrative—about cultural recognition, distributive material equality, and tools to adjudicate efforts—that we must continue to craft and refine.

REFERENCES

Bestor, Arthur E. (2008/1953). "On the Education and Certification of Teachers." In Marilyn Cochran-Smith, Sharon Feiman-Nemser, and D. John McIntyre (eds.), *Handbook of Research on Teacher Education*. New York: Routledge. 947–952.

Freire, P. (1970). *The Pedagogy of the Oppressed*. New York: Herder and Herder.

Gourx, H. (2003). "Public Pedagogy and the Politics of Resistance: Notes on a Critical Theory of Educational Struggle." *Educational Philosophy*, 35(1), 5–16.

Loughran, J. (2008). "Toward a Better Understanding of Teaching and Learning about Teaching." In Marilyn Cochran-Smith, Sharon Feiman-Nemser, and D. John McIntyre (eds.), *Handbook of Research on Teacher Education*. New York: Routledge. 1177–1182.

Popkewitz. T. (2001). "Dewey and Vygotsky: Ideas in Historical Spaces." In Thomas S. Popkewitz, Barry M. Franklin, and Miguel A. Pereyra (eds.), *Cultural History and Education: Critical Essays on Knowledge and School*. New York and London: Routledge Falmer.

Schulman, L. (1987). "Knowledge and Teaching: Foundations of the New Reform." *Harvard Educational Review*, 57(1), 1–22.

Pedagogy. Retrieved May 1, 2008, from www.joanwink.com.

CONTRIBUTORS

Michael W. Apple is John Bascom Professor of Curriculum and Instruction and Educational Policy Studies at the University of Wisconsin, Madison. He is the author of a number of award winning books on the relationship between education and differential power. Among his recent books are *Educating the "Right" Way: Markets, Standards, God, and Inequality* (2nd ed.), *Democratic Schools: Lessons in Powerful Education* (2nd ed.), with James Beane, and *The Subaltern Speak: Curriculum, Power, and Educational Struggles*, with Kristen Buras.

Wayne W. Au is an Assistant Professor in the Department of Secondary Education at California State University, Fullerton. He writes extensively on critical educational theory and social justice education sits on the editorial board for the progressive education journal, *Rethinking Schools*, and is author of the forthcoming book, *Unequal by Design: High-stakes Testing and the Standardization of Inequality*.

William (Bill) Cope is Research Professor in the Department of Educational Policy Studies at the University of Illinois, Urbana-Champaign. His current research interests include population and community diversity, theories and practices of pedagogy, and new technologies of representation and communication, including the "semantic web." For more than twenty years, Professor Cope has been a Director of Common Ground Publishing, which runs a number of annual conferences and founded the collaborative online publishing environment, CommonGroundPUBLISHER. While in Australia, he served as First Assistant Secretary in the Department of the Prime Minister and Cabinet and Director of the Office of Multicultural Affairs. His research and writing, jointly with Mary Kalantzis, has resulted in a number of books, including *The Powers of Literacy* (1993), *Productive* Diversity (1997); *A Place in the Sun: Re-Creating the Australian Way of Life* (2000); and *Multiliteracies: Literacy Learning and the Design of Social Futures* (2000).

Gustavo E. Fischman is an Associate Professor in Education Leadership and Policy Studies at Arizona State University. His areas of specialization are comparative education, gender studies in education, and the use of image-based methodologies in educational research. Dr. Fischman actively teaches and collaborates on research projects in Argentina, Brazil, Mexico, and the United States. He is the associate editor of *Education Review* and *Educational Policy Analysis Archives*, and the author of two books (*Imagining Teachers: Rethinking Teacher Education* and *Gender y La Ley y La Tierra: Historia De Un Despojo En La Tribu Mapuche De Los Toldos*), and has coedited another two (*Crisis and Hope: The Educational Hopscotch of Latin America*; and *Critical Theories, Radical Pedagogies, and Global Conflicts*) He has published numerous articles and book chapters in international journals.

Mary Q. Foote is Assistant Professor of Mathematics Education in the Department of Elementary and Early Childhood Education at Queens College- City University of New York (CUNY). Her research interests include examining the ways in which schools can utilize home and community funds of knowledge to address the educational needs of underperforming and underserved populations of students, and she also serves as a Research Associate in the Equity Studies Research Center, examining equity and diversity issues that surface as teachers discuss their own and their students' mathematical thinking.

Luis Armando Gandin is Professor of Sociology of Education in the School of Education at the Federal University of Rio Grande do Sul in Porto Alegre, Brazil, with a Ph.D. in Educational Theory from the University of Wisconsin-Madison. He is the editor of the Journal *Currículo sem Fronteiras* (Curriculum without Borders, http://www.curriculosemfronteiras.org), an open-access, refereed journal promoting academic analyses around issues of critical and emancipatory education. Professor Gandin has published several books which deal with critical pedagogy. He has also published several book chapters, and scholarly articles in Brazil, Portugal, Australia, the United States, and the UK. Professor Gandin has been researching and writing on progressive educational reform in Brazil, more particularly in the city of Porto Alegre, where the Citizen School Project of the Popular Administration has had a great impact on the lives of thousands of children living in extreme poverty.

Wangari Pauline Gichiru is a doctoral student in the department of curriculum and instruction at the University of Wisconsin-Madison. Her current research centers on U.S. teachers' perceptions of their Somali refugee students life experiences, as well as the pedagogical implications of these views. She received her B.Ed. and teaching degree from University of Nairobi in 1995, and has worked as a volunteer teacher with refugees in her home country of Kenya. She received her master's degree in Special Education from the University of Wisconsin-Eau Claire in 2004. Her research interests include refugee education, special education, and multicultural education.

Panayota Gounari is Assistant Professor of Applied Linguistics at the University of Massachusetts Boston. Her primary areas of interest include language policy and linguistic hegemony, critical discourse analysis, the role of language in social change and the construction of human agency/democratic spaces as well as the implications for critical pedagogy. Her recent publications include *The Hegemony of English* with Donaldo Macedo and Bessie Dendrinos (2003) and *The Globalization of Racism* (2006) with Donaldo Macedo. She has also published a number of articles in academic journals as well as chapters in edited books.

Carl A. Grant is Hoefs-Bascom Professor of Teacher Education in the Department of Curriculum and Instruction at the University Wisconsin-Madison. He has written or edited thirty books or monographs in multicultural education and/or teacher education, and more than 135 articles, chapters in books, and reviews. He served as President of the National Association for Multicultural Education (NAME) from 1993 to 1999; editor of *Review of Educational Research* (RER) (1996–1999); a member of the National Research Councils Committee on Assessment and Teacher Quality (1999–2001); and was earlier the chair of the AERA Publication Committee.

Diana E. Hess is an Associate Professor of Curriculum and Instruction (social studies) at the University of Wisconsin-Madison. Since 1998 she has been researching what young people learn from deliberating highly controversial political and legal issues in schools. She is currently the lead investigator of a five-year study that seeks to understand the relationship between various approaches to democratic education in schools and the actual political engagement of young people after they leave high school. Hess also researches the ideological messages embedded in high school textbooks and other curriculum. She has recently completed a study of what curricula communicate about terrorism and 9/11 and its aftermath, and scholarship about how textbooks treat *Brown v. Board of Education* was published in 2004. Hess recently completed a book on the importance of controversial issues in schools.

Mary Kalantzis is Dean of the College of Education and Professor of Curriculum and Instruction at University of Illinois,

Urbana-Champaign and former dean at the Royal Melbourne Institute of Technology (RMIT University). While Dean at RMIT, she was elected President of the Australian Council of Deans of Education. Her research crosses a number of disciplines, including history, linguistics, education, and sociology and examines the varied themes of immigration, education, ethnicity, gender, culture, leadership, and workplace change, professional learning and training, pedagogy, and literacy learning. She has managed more than 100 research and development projects in Australia, many of which involved multiliteracies, and led an international comparative study on immigration conducted for the Organization for Economic Cooperation and Development, Paris.

Triantafillia Kostouli is an Associate Professor of School Literacies at the Department of Education, Aristotle University of Thessaloniki, Greece. Her research and teaching interests lie in the areas of school and academic literacies. She is the editor of *Writing in Context(s): Textual Practices and Learning Processes in Sociocultural Settings*, published by Springer.

Douglas B. Larkin is a doctoral student in the Department of Curriculum and Instruction at the University of Wisconsin-Madison. His current research concerns preservice teacher learning in teacher education programs, particularly in regard to the preparation of science teachers for racially and ethnically diverse classrooms. He received his B.S. and teaching degree from Trenton State College in 1993, and he served as a high school science teacher for ten years, including service as a Peace Corps Volunteer in Kenya and Papua New Guinea. He received his master's degree in multicultural science education from the University of Wisconsin-Madison in 2001. His research interests include teacher education, science education, and multicultural education.

Donaldo Macedo is professor of English and Distinguished Professor of Liberal Arts and Education at the University of Massachusetts Boston. He has published extensively in the areas of linguistics, critical literacy, and multicultural education. His publications include: *Literacy: Reading the Word and the World* (with Paulo Freire, 1987), *Literacies of Power: What Americans Are Not Allowed to Know* (1994), *Dancing with Bigotry* (with Lilia Bartolomé, 1999), *Critical Education in the New Information Age* (with Paulo Freire, Henry Giroux, and Paul Willis, 1999), *Chomsky on Miseducation* (with Noam Chomsky, 2000), *The Hegemony of English* (with Panayota Gounari and Bessie Dendrinos, 2003), *Howard Zinn on Democratic*

Education (with Howard Zinn, 2005), *The Globalization of Racism* (with Panayota Gounari, 2005), and *Media Literacy* (with Shirley Steinberg, 2007).

Glenda Mac Naughton is Professor of Learning and Educational Development and Director of the Centre for Equity and Innovation at the University of Melbourne. Her primary areas of research focus on social justice and equity issues in early childhood, curriculum development and innovation, and teacher change and critical thinking in early childhood. Author of many internationally recognized books and articles, including *Doing Foucault in Early Childhood* and *Rethinking Gender in Early Childhood Education*, Professor Mac Naughton is a frequent keynote speaker at international conferences and consultant on children's rights and participation, as well as on diversity and equity issues.

Kathryn D. Manuelito is from the Diné (Navajo) Nation and is Naakai Diné'é and born for the Kinlichiinii clan. Her research focus includes language and literacy, teacher education, self-determination in indigenous communities, decolonizing methodologies, and indigenous womanism. She is an associate professor at the University of New Mexico after serving as chair of Indian Education at Arizona State University.

Mónica Mazzolo is the Director of *Escuela Pública* 10 (Public School 10) in Luján, Argentina. She has developed a curriculum at the school that is integrated and project based. This curriculum was expanded in important ways to address the underperformance of Bolivian immigrant children in the school.

Soula Mitakidou is an Associate Professor in the Department of Primary Education at Aristotle University of Thessaloniki, Greece, where she teaches antiracist education and second language acquisition among nonnative speakers. Her recent publications and presentations focus on many aspects of diversity, including integrated instructional strategies and marginalized learners.

Bekisizwe S. Ndimande is Assistant Professor of Curriculum and Instruction and with the Center for African Studies at the University of Illinois at Urbana-Champaign. His research interests include the politics of curriculum and examining the policies and practices in postapartheid desegregated public schools, and also the implications of school "choice" for disadvantaged communities in South Africa. His current research examines children's rights and experiences of immigrant students and families in South Africa.

Maxine Roanhorse-Dineyazhe is originally from Chinle, Arizona, on the Navajo Nation. Inspired by her son's educational experiences, she pursued a degree in education. While teaching elementary school in various American Indian communities, Maxine began graduate studies and earned her Master's degree in 1997. In 2007, Maxine completed her doctoral research on the recruitment factors of American Indian teacher education candidates and earned her doctorate degree at Arizona State University. Maxine served as Program Coordinator for the Indigenous Teacher Preparation Program at Arizona State University and is currently serving as Director of Curriculum and Instruction for Shonto Preparatory School on the Navajo Nation. According to her, "The importance of learning & education is reflected in my grandmother's teachings, guided by my mother's examples, and nurtured in my son's future."

Alberto J. Rodriguez is Associate Professor of Education at San Diego State University, where he teaches courses in the graduate program, and also teaches science methods courses in the bilingual teacher credential program. His research focuses on the use of sociotransformative constructivism (sTc) to help teachers teach for understanding in diverse school contexts. sTc is a theoretical framework that merges multicultural education (a theory of social justice) with social constructivism (a theory of learning). He has authored a number of journal publications and is coeditor of a new book, *Managing Teachers' Resistance to Teach for Diversity and Understanding: Strategies for Transformative Action*.

Christine E. Sleeter is Professor Emerita in the College of Professional Studies at California State University, Monterey Bay. She was recently a Visiting Professor at Victoria University in New Zealand and at the University of Washington, Seattle, and she served as Vice President of Division K (Teaching and Teacher Education) of the American Educational Research Association. Her research focuses on antiracist multicultural education and multicultural teacher education. Her most recent books include *Un-Standardizing Curriculum* (Teachers College Press), *Facing Accountability in Education* (Teachers College Press), and *Doing Multicultural Education for Achievement and Equity* (with Carl Grant; Routledge).

Lourdes Diaz Soto is Goizueta Endowed Chair and Professor of Education at Dalton State College in Dalton, Georgia. Some of her publications include The Praeger *Handbook of Latino Education in the U.S*, *Making a Difference in the lives of Bilingual/Bicultural Children, and Power and Voice in Research with Children*, and many journal articles and book chapters. She teaches critical pedagogy and bilingual education and is currently conducting research with her graduate students viewing how issues of social justice and equity impact Latina/os in the United States.

Beth Blue Swadener is Professor and Chair of Early Childhood Education and Professor of Policy Studies at Arizona State University. Her research focuses on social policy, anti-oppressive strategies in early childhood contexts, child and family policies in sub-Saharan Africa, and transnational comparative research on children's rights and participation. She has published nine books, including *Children and Families "At Promise": Deconstructing the Discourse of Risk, Does the Village Still Raise the Child?: A Collaborative Study of Changing Childrearing and Early Education in Kenya, Decolonizing Research in Cross-Cultural Context, and Power and Voice in Research with Children*, and numerous articles and book chapters.

Evangelia Tressou is a Professor of Pedagogy with emphasis on the education of special groups and on mathematics teaching at the Primary Education Department of Aristotle University of Thessaloniki, Greece. Her main research and teaching interests focus on the education of linguistic and cultural minorities, the problems they face with mathematics, as well as the relation of gender with mathematics.

Joseph Tobin is Basha Professor of Early Childhood Education at Arizona State University. His research interests include cross-cultural studies of early childhood education, immigration and education, children and the media, and qualitative research methods. Among his publications are *Preschool in Three Cultures, Good Guys Don't Wear Hats: Children's Talk about the Media, Making a Place for Pleasure in Early Childhood Education, and Pikachu's Global Adventure: The Rise and Fall of Pokemon*. He is currently completing a sequel to *Preschool in Three Cultures* and directing the Children Crossing Borders project, a five-country study of parent and staff views about what children of recent immigrants should experience in ECEC settings.

Shannon Murto Wright earned a Master of Science degree in Curriculum and Instruction from the University of Wisconsin-Madison in 2006. Her research interests include democratic education, controversial issues teaching, and standardized testing in the social studies.

INDEX